The Matter of the Mind

The Matter of the Mind

Philosophical Essays on Psychology, Neuroscience, and Reduction

Edited by Maurice Schouten
and Huib Looren de Jong

Blackwell
Publishing

Parts of Chapter 10 were published earlier in Andy Clark, "Happy Couplings: Emergence and Explanatory Interlock," from Margaret A. Boden (ed.), *The Philosophy of Artificial Life* (Oxford: Oxford University Press, 1996). By permission of Oxford University Press.

BLACKWELL PUBLISHING
350 Main Street, Malden, MA 02148-5020, USA
9600 Garsington Road, Oxford OX4 2DQ, UK
550 Swanston Street, Carlton, Victoria 3053, Australia

First published 2007 by Blackwell Publishing Ltd

1 2007

Library of Congress Cataloging-in-Publication Data

The matter of the mind: philosophical essays on psychology, neuroscience, and reduction / edited by Maurice Schouten and Huib Looren de Jong.
p. cm.
Includes bibliographical references and index.
ISBN-13: 978-1-4051-4443-8 (hardback : alk. paper)
ISBN-10: 1-4051-4443-2 (hardback : alk. paper)
1. Reductionism. 2. Neurosciences—Philosophy. 3. Psychology—Philosophy. I. Schouten, Maurice Kenneth Davy, 1970– II. Looren de Jong, Huibert.
B835.5.M38 2007
128′.2—dc22
2006021667

A catalogue record for this title is available from the British Library.

Set in 10.5/13pt Galliard
by Graphicraft Limited, Hong Kong
Printed and bound in United Kingdom
by TJ International Ltd, Padstow, Cornwall

The publisher's policy is to use permanent paper from mills that operate a sustainable forestry policy, and which has been manufactured from pulp processed using acid-free and elementary chlorine-free practices. Furthermore, the publisher ensures that the text paper and cover board used have met acceptable environmental accreditation standards.

For further information on
Blackwell Publishing, visit our website:
www.blackwellpublishing.com

CONTENTS

CONTRIBUTORS

Bechtel, William
Department of Philosophy, Science Studies Program, and Interdisciplinary Program in Cognitive Science, University of California, San Diego, 9500 Gilman Drive, LaJolla, CA 92093, USA. E-mail: bill@mechanism.ucsd.edu

Bickle, John
Department of Philosophy and Neuroscience Graduate Program, University of Cincinnati, PO Box 210374, Cincinnati, OH 45221-0374, USA. E-mail: bicklejw@email.uc.edu

Clark, Andy
Department of Philosophy, University of Edinburgh, George Square, Edinburgh, EH8 9JX, Scotland. E-mail: Andy.Clark@ed.ac.uk

Endicott, Ronald
Department of Philosophy and Religion, College of Humanities and Social Sciences, North Carolina State University, Campus Box 8103, Raleigh, NC 27695-8103, USA. E-mail: rpendico@ncsu.edu

Gillett, Carl
Department of Philosophy, Illinois Wesleyan University, Bloomington, IL 61702-2900, USA. E-mail: cgillett@iwu.edu

Looren de Jong, Huib
Department of Psychology and Faculty of Philosophy, Vrije Universiteit, Van der Boechorststraat 1, 1081 BT Amsterdam, The Netherlands. E-mail: H.Looren.de.Jong@psy.vu.nl

McCauley, Robert N.
Department of Philosophy, 561 S. Kilgo Circle, Emory University, Atlanta, GA 30322, USA. E-mail: philrnm@emory.edu

Melnyk, Andrew
Department of Philosophy, University of Missouri-Columbia, 438 General Classroom Building, Columbia, MO 65211-4160, USA. E-mail: melnyka@missouri.edu

Polger, Thomas W.
Department of Philosophy, University of Cincinnati, PO Box 210374, Cincinnati, OH 45221-0374, USA. E-mail: thomas.polger@uc.edu

Richardson, Robert C.
Department of Philosophy, University of Cincinnati, PO Box 210374, Cincinnati, OH 45221-0374, USA. E-mail: robert.richardson@uc.edu

Schouten, Maurice K. D.
Faculty of Philosophy, Tilburg University, Warandelaan 2, PO Box 90153, 5000 LE, Tilburg, The Netherlands. E-mail: m.k.d.schouten@uvt.nl

Shapiro, Lawrence A.
Department of Philosophy, University of Wisconsin, 5185 Helen C. White Hall, Madison, WI 53706, USA. E-mail: lshapiro@wisc.edu

Wright, Cory D.
Department of Philosophy, University of California, San Diego, 9500 Gilman Drive, La Jolla, CA 92093-0119 USA. E-mail: sarahbellum@mad.scientist.com

PREFACE AND ACKNOWLEDGMENTS

Like Caesar's Gaul, this book is divided into three parts, and like Caesar's book (*The Conquest of Gaul*), it is about domains, border conflicts, and imperialism – in this case, the shifting border between psychology and neuroscience, and the possibility that psychology will be annexed by and incorporated in neuroscience.

Is there a possibility that psychology will be reduced to or even replaced by neuroscience? That depends on what is meant by reduction in general (Part I), it depends on how theories in different sciences can be related (Part II), and it depends on empirical evidence (Part III). Thus, in this book, both the philosophical framework, the conceptual and metaphysical foundations, as well as the empirical evidence for such reductive claims are addressed.

In Part I the ontological framework underlying reduction claims is explored: what does it mean that a higher-level phenomenon or property is reduced to a lower level? On the basis of recent work in the metaphysics of science, the conceptual geography and ontological commitments are investigated. In Part II a subtly different question is addressed: even if mental processes are essentially brain processes, what does that mean for theory building in psychology and cognitive science? Philosophers of science trace the dauntingly complex relations between scientific theories, and analyze how lower-level (neural) mechanisms can be said to explain or reduce higher-level (mental) functions. In Part III the empirical evidence for potential reductions is assessed: is it possible to reduce psychological phenomena as emergent adaptive behavior, consciousness, desire, and reward, and empathy in understanding others, to brains, or even to nerve cells?

Until recently an antireductionist consensus was firmly in place in philosophy: psychology was supposed to remain (relatively) autonomous with respect to the sciences of the brain. As the chapters in this book show, things are changing. Reductionism has become fashionable again. Cutting-edge developments in neuroscience have put the issue of the status of psychology – and its relation to neuroscience – firmly back on the agenda in both science and philosophy.

Opinions among the authors in this book range from uncompromising reductionism to explanatory pluralism. Their philosophical styles range from conceptual analysis, via philosophical reflection on the nature of scientific explanation, to metascientific interpretation of laboratory data. Some say that interesting psychoneural reductions have as a matter of fact already been accomplished. Others see continuing top-down influences from psychology to neuroscience. Together, they present a more complex, rich, and sophisticated picture of reduction than the simple dichotomy between reductionist and antireductionist. The interface of psychology and neuroscience may sometimes look like the scene of a border conflict, but it is a dynamic trade zone as well.

We would like to express our gratitude to the authors who contributed to this volume. It was a pleasure to work with them. In addition, we want to thank Bill Bechtel, John Bickle, Dingmar van Eck, and Cory Wright for help and support, good ideas and useful suggestions, valuable comments, and some very speedy reviews.

Maurice Schouten and Huib Looren de Jong
Tilburg and Amsterdam

1

MIND MATTERS: THE ROOTS OF REDUCTIONISM

Maurice Schouten and Huib Looren de Jong

1. Introduction

How are the mental and the physical related? The question concerning the *commercium mentis et corporis* has troubled scientists and philosophers for ages. Descartes's solution in terms of a dualism of substances, interacting at the *conarion*, is now considered a relic of a very distant past. Science and philosophy have turned materialist: all that exists, exists in space and time and must be considered fundamentally physical.

Though one may be convinced that we inhabit a universe that is materially constituted, the question remains whether such an ontological physicalism at the same time commits one to reductionism: Are minds nothing but brains? Will, when all is said and done, psychology really be nothing more than a chapter in neuroscience? Oppenheim and Putnam claimed as much in 1958 when they suggested that "[i]t is not absurd to suppose that psychological laws may eventually be explained in terms of the behavior of individual neurons in the brain" (p. 7). Still, Putnam himself was to a considerable extent responsible for the firm "antireductionistic consensus" that emerged in the philosophy of science and the philosophy of mind. According to the ruling orthodoxy, mainly due to Putnam and Fodor's "multiple realizability" argument (Putnam, 1960; Fodor,

1975), reductionism cannot possibly be true. The fact that mental functions can be instantiated in a wide variety of underlying physical substrates precludes them from being reductively mapped onto neurophysiological processes. It is in this antireductionistic climate that "reductionism" became a term often used with pejorative intent: "a general term of insult and abuse" (Churchland, 1986, p. 278), "a dirty word" (Dawkins, 1982, p. 113), a term that refers to something "philistine and heartless, if not downright evil" (Dennett, 1995, p. 80). Most philosophers of mind have opted for nonreductive forms of physicalism: mental properties are not identical to physical properties and psychology will continue to enjoy autonomy relative to the neurosciences. Anyone who claims otherwise must be considered "an imperialist in the service of physics" (Brooks, 1994, p. 803).

These days however, it surely looks like the pendulum is swinging back to reductionism again. The writ of reductionism has been spreading across the sciences, and its effects on our views of the world are pervasive. The ever-increasing momentum with which in modern neuroscience and molecular biology discoveries are made and theories are formulated reinvigorates reductionist claims with respect to the traditional territory of psychology. Cracks have begun to appear in the apparently solid phalanx of support for the autonomous status of psychology. The availability of intimate correlations between psychological phenomena and neurophysiological activity has sparked a renaissance of reductionism.

In this introductory chapter, we will take reduction to be about the relations between levels: between levels of description and explanation, or between levels of reality. Higher-level explanations seem threatened in two ways, in a Catch-22-like fashion: damned when fitting in a physical world, and damned if they don't. On the one hand, when higher-level posits cannot be related to the real furniture of the world, as captured in the laws of macrophysics, they can't be real things or processes in a causally closed world, can't really explain anything. On the other hand, if a higher-level explanation can be related to physical processes, it becomes redundant, since the explanatory work can then be done by physics. So, in exploring reduction the crucial point is whether and how higher levels (biology, psychology, and their objects) connect to more basic levels of reality and explanation.

The crucial questions are whether the entities at higher levels (such as societies, minds, and adaptive functions) have a reality unto themselves and whether the theories or domains of enquiry (such as sociology, psychology, and biology) that try to describe and explain them exhibit conceptual integrity or provide genuine explanations. Presumably, the lower-level (roughly macrophysics) is usually considered unproblematic: hardly anyone ever seems to question the conceptual integrity, explanatory power, and reality of physics. Intentional explanations in psychology and functional explanations in biology are under constant threat of being replaced by lower-level explanations.

Although we may feel uncomfortable with such conclusions, this may just be the road that lies ahead. As E. O. Wilson states: "reductionism is the primary and essential activity of science" (Wilson, 1998, p. 54). Klein & Lachièze-Rey agree when they say that "[s]cience is reductionistic by essence" (1999, p. 129). Reduction is essential in science because – as is often claimed – nature (and our views of nature) must be unified. One reason that has always motivated reductionist projects is the appeal to Occam's Razor or *ontological simplicity.* The "nothing-buttery" locution is the reductionist's battle cry. Accomplished reductions leave us with fewer entities in our catalog of the universe. In case one would be able to show that mental events are neural events, one would have a more parsimonious ontology. Another important motivation for reductionism is *explanatory parsimony*: a reduction leaves one with theories that are more comprehensive and more predictively and explanatorily powerful than the ones had before. As Otto Neurath, in the introductory essay to the *International Encyclopedia of Unified Science*, stated: "All-embracing vision and thought is an old desire of humanity" (cited in Suppes, 1981, p. 3). The ideal of a *Einheitswissenschaft* was not only central to the Vienna Circle positivists; the desire to integrate disparate pieces of knowledge can be found in, to name but a few, Francis Bacon, Descartes, Kant, and Leibniz. Leibniz asserted that "The entire body of the sciences may be regarded as an ocean, continuous everywhere and without a break or division, though men conceive parts in it and give their names according to their convenience." Similar ideas can be found in Kant, who wrote that "our diverse modes of knowledge must not be permitted to be a mere rhapsody, but must form a system" and "Every science is a system in its own right; . . . we must . . . set to work architectonically with it as a self-subsisting whole, and not as a wing or section of another building – although we may subsequently make a passage to or from one part to another" (citations in McRae, 1957,

p. 1). However, this regulative idea of integrating science was codified into a rigid, formalized prescription for unification through reduction by Ernest Nagel (1961). This classical view of reduction is more like seizing a neighbor's property and rebuilding it, than like making passages between domains of knowledge.

2. Classical Reductionism and the Problem of Connectability

2.1 Classical Reduction

"[T]he phenomenon of a relatively autonomous theory becoming absorbed by, or reduced to, some other more inclusive theory is an undeniable and recurrent feature of the history of modern science," writes Nagel in his *locus classicus*, *The Structure of Science* (Nagel, 1961, pp. 336–337). What is required to effect such a reduction? On Nagel's account, reduction involves a relation between two scientific theories, a secondary or target theory T_R and a primary or successor theory T_B. In essence, according to Nagel, the satisfaction of two conditions forms the key to a successful reduction of one theory to another. The first is the "condition of derivability" [DC]: reduction is essentially a matter of the logical derivation of T_R from T_B, for instance, the derivation of thermodynamics (specifically, the Boyle–Charles law) from statistical mechanics (plus the kinetic theory). The second condition that must be fulfilled – what Nagel called the "condition of connectability" (Nagel, 1961, p. 354; [CC]) – becomes especially clear when one considers *heterogeneous* theory connections, i.e., cases in which the proprietary vocabularies of the primary and secondary theory show no (full) overlap. For a deductive argument to be valid it is required that T_B is supplemented by statements that connect the terms that occur in its laws and postulates and those terms which are peculiar to T_R. So, for instance, "temperature" does not occur in statistical mechanics and should be correlated with "mean molecular energy," one of T_B's proprietary terms.

In case both [CC] and [DC] are satisfied, the old, secondary theory (or something similar to it) is incorporated by the new (primary) one and comes out as a special case of the new theory under limited conditions. For example, Kepler's laws are reduced by subsuming them under Newton's laws of motion and gravitation; and so the former set of laws is a special case of the latter set.

2.2 Strong Connectability

On Nagel's canonical view of reduction, it is demanded that the vocabularies of T_R and T_B be correlated via bridge laws, but what is the status of these bridging principles? For Nagel, the bridging principles were universally quantified biconditionals or even one-way conditionals (Nagel, 1961, p. 355n). However, numerous commentators have pointed out that this is surely too weak (Causey, 1972; Enç, 1976). As Hooker explains: "Nagel's conditions . . . are too weak to ensure the dispensability of either the reduced theory's conceptual apparatus or its ontology" (Hooker, 1981, p. 39). We need something stronger than mere correlations, because – although T_R would be derivable from T_B – one would be faced with the further task of explaining the correlation laws. An additional (bridge) theory is needed that explains the correlations between T_R and T_B. This implies that with the fulfillment of [CC], [DC] can indeed be satisfied quite easily, however not in the way Nagel envisaged. T_R is derivable, not from T_B alone, but from a conjunction of T_B and a set of correlatory statements and, concludes Sklar, "this reduction is not the reduction of [T_R] to [T_B] originally sought for" (1967, p. 119).

Nagelian reductions fail to make good on the promise to shrink ontologies and vocabularies. Establishing correlations between theories T_B and T_R are sufficient for derivation of T_R from T_B, but it is a mistake to claim – as Nagel did – that mere correlations will result in a reduction of T_R to T_B. Neither *explanatory* nor *ontological economy*, the principal motivating aims behind reductionism, will have increased. In the words of Kim:

> By adding the bridge laws to the reductive resources as auxiliary premises, *Nagel reduction essentially extends the reduction base*. If we take reduction to be an explanatory process which yields an explanation of the laws and phenomena being reduced on the basis of the laws of the base theory, Nagel reduction fails to generate such explanations. For, to do so, the reductive derivation must derive the laws being reduced *solely from the explanatory resources available in the base domain*." (Kim, 2005, pp. 99–100; original emphasis)

Moreover, Nagelian reduction fails to deliver *ontological parsimony* as well. As a number of commentators have insisted, mere correlations can never succeed in ontological economizing since the ontologies of T_R and T_B remain distinct: to say that mental states and neural processes are correlated is "to say that they are something 'over and above.' You cannot correlate something with itself" (Smart, 1959, p. 142).

In light of these considerations, it looks as if Nagel's model is in need of modification, and many have suggested that a good place to start is [CC]. Whereas Nagel was at pains to avoid ontological commitments, it is debatable whether his neutral stance on matters ontological can be sustained. On close inspection, it looks like additional restrictions need to be imposed upon the postulated bridge laws to get the reductionist's project off the ground. In particular, Sklar (1967) argued that the only way to get rid of the correlatory statements that connect *two* classes of entities – which are themselves in need of further explanation – is to demand that T_R and T_B are (strongly) connected through empirically established *identity* statements (Schaffner, 1967, p. 144; Sklar, 1967, p. 120). Without identities strongly connecting T_R and T_B, "the underlying ontological bias of the reductionist program" would not be satisfied as mere correlations are "compatible with a nonphysicalist ontology" (Fodor 1974, p. 129). Kims dubs this identity requirement the "condition of strong connectability" (Kim, 1993, p. 151). We will refer to this condition as [sCC].

2.3 The Failure of Connectability I: Multiple Realization

What are the prospects of psychoneural reductionism? Only on a strong reading of bridge laws ([sCC]) can the classical reductionist's program be rendered truly successful in the psychoneural case. Hence, early psychoneural reductionists defended a Psychoneural Identity Theory (Feigl, 1958; Place, 1956; Smart, 1959). They claimed that mental states and events can be empirically identified with neural states and events just as lightning can be identified with electric discharges. In short, mental kinds are nothing but neural kinds. However, here the Psychoneural Identity Theory immediately faces the *correlation objection* (cf. McCauley and Bechtel, 2001). Brandt and Kim (1967) formulated this objection against the logic of the identity theory thus: "since the identity statements have no more empirically verifiable content than their associated correlations, the theory with identities will fare as well as the theory with correlation laws, in confrontation with observational fact" (Brandt & Kim, 1967, p. 530). Briefly, finding mind-brain covariances is not enough to support mind–brain identities.

Besides this objection, there were other influential lines of argumentation directed against the view that there are strong (identity) connections between the mental and the physical, e.g., Davidson's *anomalous monism*. Davidson argued that unlike physical events, mental events are not governed by strict laws; hence, there are no nomological connections between the mental and the physical. Another important pressure source is the well-worn multiple realizability argument, first formulated by Putnam and later generalized by Fodor. Putnam claimed that mental states can be and typically are implemented by many, wildly diverse physical states. This makes the implementation level explanatorily uninteresting. In Putnam's happy phrase, "We could be made of Swiss cheese and it wouldn't matter" (Putnam, 1975, p. 291). Moreover, the fact that mental functions can be instantiated in a wide variety of material substrates precludes them from being reductively mapped onto, say, neurophysiological processes. Again, for the reductionist program in psychology to succeed, psychological kind predicates should be lawfully coextensive with neural kind predicates, but they are not. Hence, given multiple realizability, psychoneural reductionism must be ruled out: "what corresponds to the kind predicates of a reduced science may be a heterogeneous and unsystematic disjunction of predicates in the reducing science" (Fodor, 1981). Hence, Fodor's conclusion is that multiple realizability "refutes psychophysical reductionism once and for all" (Fodor, 1998, p. 9).

2.4 The Failure of Connectability II: Approximation, Correction, Radical Falsity

The very soundness of the classical model of reduction has been disputed in other quarters of philosophy as well. One objection often raised against Nagel's treatment of theory reduction is that it fares badly in terms of historical accuracy (Caplan, 1981). There are few – if any – cases of intertheoretic relations that qualify as reductions on this view. The most quoted example is the reduction of classical thermodynamics to statistical mechanics supplemented by the kinetic theory of matter. The classical model apparently not only fails as an account of the psychology-neuroscience case; nor can it explain the relation between classical genetics and molecular genetics (Hull, 1974). Can the classical model be remedied so as to provide a better fit with scientific history? A number of authors have thought so.

Let us start with what is commonly seen in the philosophy of science literature as the main failure of classical reductionism, the fact that it disregards incompatibilities or discrepancies between T_R and T_B and that it fails to account for the possibility of correction of the theory targeted for reduction (see Kemeny & Oppenheim, 1956; Popper, 1957; Sellars, 1965). Popper claimed that "from a logical point of view, Newton's theory, strictly speaking, contradicts both Galileo's and Kepler's" (Popper, 1957, pp. 29–30). For instance, for Galileo – contradicting Aristotelian physics – a stone that is thrown follows a parabolic trajectory, whereas for Newton – contradicting Galileo – the path of the stone will be elliptic. The trajectory becomes *approximately* a parabola only at relatively small distances. Similar things can be said about the alleged reduction of Kepler's laws to Newton's mechanics so that "Kepler's laws are only approximately valid" (Popper, 1957, p. 32). Sklar (1967) is clear on the implication: "even in the case of homogeneous theories reduction is very rarely derivation" (p. 2). Although in general sympathetic to Nagel's project, Hempel diagnosed the received (Nagelian) view of reductionism to be an "untenable oversimplification which has no strict application in science and which, moreover, conceals some highly important aspects of the relationship to be analyzed" (Hempel, 1969, p. 197). The conclusion seems warranted that Nagelian reduction is just an empty formalism, an idealization at best (in fact, this is consistent with Nagel's characterization of reduction as laid out in his 1961 as an "ideal demand" – see p. 347).

Prompted by the critics of the received view of reduction, Nagel, Hempel, and Kenneth Schaffner have attempted to handle these objections. What they claimed is that by introducing a notion of *approximative reduction*, the incompatibilities between T_R and T_B could be accounted for (Gaa, 1975). As Nagel recognizes in response to some of his critics (and speaking of homogeneous reductions): "the laws derivable from Newtonian theory do not coincide exactly with some of the previously entertained hypotheses about the motions of bodies" (Nagel, 1970, p. 120), however "the initial hypotheses [T_R] may be reasonably close approximations to the consequences entailed by the comprehensive theory [T_B]" (p. 121). What is deduced from T_B is not T_R but "the *approximate* truth of the reduced theory" (Putnam, 1965, p. 206).

As Schaffner (1967) points out, in case some new lower-level theory T_B shows up, T_R is often revised into a theory T_R^* which stands in a relation of "strong analogy" to T_R and in which T_R's false elements are removed. For example, statistical mechanics redefines "temperature" (Brittan, 1970); what has taken place is not a reduction of classical thermodynamics, as the received view pictured the intertheoretic relation, but "something resembling it" (p. 453). Thus, Nagel's [DC] remains in force, but now the entailment holds between T_B and an appropriate, "strongly analogous" image T_R^* of T_R (Schaffner, 1967; 1974), a "corrected secondary theory" (Schaffner, 1967). According to Schaffner, one will now be able to bypass the incompatibilities between T_R and T_B because these can be removed in T_R's approximative image T_R^*. As Gaa succinctly formulates it: "The condition of derivability, so important to Nagel, now requires, for the relation of reduction to hold between two theories, that an *appropriate* analog of the reduced theory, and *not* the reduced theory itself, be derived from the reducing theory" (Gaa, 1975, p. 355).

Many problems with this notion of approximative reduction have been pointed out. What may count as an adequate approximation to T_R? Feyerabend claimed that the relation between T_R and T_R^* is "too vague" and "essentially subjective" (1981, pp. 58–59). The real problem is that "real theories, theories which have been discussed in the scientific literature, are replaced by emasculated caricatures" (Feyerabend, 1965, p. 229). In actual science, Kuhn and Feyerabend argued, there are many cases in which T_R and T_B are incommensurable. Cases of theory change often violate what Feyerabend termed the "condition of meaning invariance" and this implies that T_R/T_R^* cannot simply be derived from T_B. What one often observes in science are revolutions rather than the cumulative and progressive change envisioned by Nagel and his followers.

Time to take stock: Nagel's elegant account of theory reduction does not work. First, the world does not cooperate: the kinds of psychology (representations, consciousness, qualia) do not have neat bridge law-like connections with the kinds of physiology or physics (the same applies to biology, for that matter). Second, science does not conform to this model: with the progress of science, meanings change and old theories are rewritten (sometimes beyond recognition) rather than smoothly incorporated in the new theory.

3. The New Reductionisms

3.1 New Wave Reductionism

In science, one is often confronted with what may be termed "replacement reductions" (Sklar, 1967, p. 4). Hooker even claimed that in fact one of Nagel's cherished examples of a theory reduction may be such a replacement reduction: "thermodynamics is simply conceptually and empirically wrong and must be replaced" (Hooker, 1981, p. 49). In cases of replacement, (strong) bridge laws or reduction functions are obviously not obtainable: one cannot formulate identity statements if at least one of the terms is referentially empty. Applicability in science being an important desideratum for any model of theory change, the question becomes: how to account for T_Rs which are *radically* false?

Schaffner's (1967) General Reduction Paradigm, later remodeled into the General Reduction-Replacement Paradigm (GRR), was the first attempt at a formal rewrite of Nagel's model by weakening [CC] and [DC]. GRR aimed to reconcile the seemingly incompatible views of scientific change, i.e., Nagel's account of "smooth" theory changes and Kuhn/Feyerabend's view of "bumpy" theory changes, of scientific progress might be reconciled in one comprehensive model. Even when Nagel's [DC] cannot be met, in particular in cases that involve false theories, one may still be able to map theories on one another. New Wave Reductionism is based on Schaffner's "General Reduction-Replacement model"; however, it considers Schaffner's view way too liberal because the latter allows T_R^* to be built out of materials supplied by the uncorrected T_R and this theory may be completely mistaken. In contrast, NWR demands that a corrected image or analog of T_R is constructed *out of the conceptual resources furnished by* T_B (Hooker, 1981, p. 49). This analog, T_R^*, mimics to some extent the formal/structural properties of T_R. T_R^* is an "analog" *within* T_B. It is this appropriately revised version of T_R in the base theory T_B which is derived, not the laws of T_R themselves: "what is explained directly by the reducing theory are the *corrected* statements derivable from it" (Hooker, 1981, p. 46). Thus, deduction on the "new wave" model is always an *intra*theoretic relation, not an intertheoretic one, as in Nagel reductionism. This feature allows the model to account not only for reduction cases in which T_R's referents are retained, but also for cases of elimination (Bickle, 1998, p. 29; see also Schouten & Looren de Jong, 1999).

With NWR, demands with respect to the relation between T_R and T_R^* are less stringent than [CC] and [sCC]. Only a relation of analogy is required, not a strict identity expressed in a bridge law. This feature allows the obtained intertheoretic analogies to be ordered along a dimension of "perfectly smooth" cases (*retentive reductions*) to "extremely bumpy" cases (*eliminative reductions*). Whenever the mapping of T_R onto T_R^* is subject only to comparably minor revisions, the mapping is smooth. This means that a reduction has been achieved, which implies that the ontology of T_R is preserved. In those instances in which T_R and T_R^* are relatively or even radically dissimilar, however, because large-scale revisions were necessary to construe T_R^*, an elimination of (parts of) T_R will have obtained. In particular by loosening Nagel's [CC], NWR is able to sidestep many of the problems that troubled *ancien régime* reductionism. It is by dropping [CC] (and [sCC] for that matter) that NWR accommodates eliminative reductions. More specifically, it accounts for the possibility of (folk) psychology being eliminated by neuroscience.

Eliminative materialism was brought into stark relief by Paul Churchland. It asserts that folk psychology is a relic of the past, hopelessly disconnected from the rest of the scientific world. Churchland's diagnosis is that "Folk Psychology is a modern cousin of an old friend: Ptolemaic Astronomy" (Churchland, 2005, p. 38) and its items belong in the museum of antiquities along with such curiosities as entelechies, élan vital, crystal spheres, phlogiston, ether, witchcraft, sunrises, and so on, which have all been displaced from our best scientific ontologies. In terms of NWR this means that the (folk)psychology-neuroscience case falls at the eliminative end of the continuum of intertheoretic analogies.

3.2 Functional Reductionism

Recall Fodor's remark that the multiple realizability argument "refutes psychophysical reductionism once and for all" (1998, p. 9). Many in the philosophy of mind followed Fodor and embraced the multiple realizability argument as a "Declaration of Independence" (Shapiro, Chapter 5, this volume), as it apparently succeeded in securing a robust autonomy for the mental *vis-à-vis* the neurophysiological. In recent years, however, many authors have expressed doubts concerning the force of the argument (Bechtel & Mundale, 1999; Bickle, 1998; Churchland, 1986; Enç, 1983; Polger, 2004; Shapiro, 2004). According to Churchland, for instance, multiple realizability can not be an obstacle to intertheoretic reduction since multiple realizability even obtains in such a textbook case of intertheoretic identification (and reduction) as temperature being identical to mean molecular kinetic energy (discussed in Bickle, 1998). Whereas temperature in a gas is identical to mean molecular kinetic energy, temperature in a solid or in a plasma is not. Moreover, if multiple realizability really were a problem, it is hard to explain the current successes of neurosciences, as these are built on the premise that there is genuine continuity of function across individuals and even across species. Even if multiple realizability is real, it may not be able to block intertheoretic reduction. Even if "global" bridge laws are unavailable, "species- or structure-specific bridge laws" remain a possibility; while we may have to give up on "global" (Nagel-style) reductionism, one might still have "local reductions" and therefore psychology can no longer enjoy autonomy (Kim, 1998).

Kim's (1998, 2005) metaphysical work illustrates the physicalist Catch-22 mentioned earlier: if mind fits in a physical world, it exists only in virtue of a physical realization; if it does not fit in the physical world, it cannot be real. Saving mental causation (and by analogy, higher-level explanations) consists in showing its physical realization. This, briefly, is how the argument is supposed to undercut nonreductive physicalism.

The starting-point for Kim is nonreductive physicalism's commitment to the view that mental properties have new causal powers over and above the causal powers of neural, or other physical, properties. When I want to have a beer and I think I can have one by opening the fridge, I walk up there because I have this desire and this belief – and not, say, because my neurons in this or that part of my central nervous system are firing. But what relation between mental and neural properties justifies the nonreductionist in his claim that it was my belief and my desire that caused my behavior and not the activity of my central nervous system, especially given the nonreductionist's commitment to physicalism. If this relation is not identity, then what is it?

It is a familiar story that nonreductive physicalists have turned to supervenience: mental properties are dependent on, or determined by, physical properties, but physical properties are not dependent on, or determined by, physical properties. It is this relation of asymmetrical dependence that is supposed to safeguard the autonomy of the mental.

According to Kim, there are good reasons to doubt that supervenience will be able to make room for the mental in a physical world. Under the nonreductivist's assumption of supervenience, a mental property *M* is causally active because a neural property *P* is. For instance, I have a desire about a bottle of my favorite Trappist beer in the fridge.[1] This desire depends – by supervenience – on a specific neural state. So, what causes me to open my Westmalle Dubbel? Although there is a causal connection between mental property *M*, my desire, and me pouring myself a glass of beer, the latter action was also causally necessitated by the physical (presumably neural) property *P* on which mental property *M* supervenes. So, how can *M* *really* be a cause of my action when I have *M* because I have *P*? What causal work is there to do for *M* over and above the work already carried out by *P*? Hence, any causal story involving mental states like beliefs and desires will be pre-empted or excluded by a more fundamental neurophysiological story and the nonreductivist's claim to autonomous mental efficacy is unjustified. (See Shapiro, Chapter 5, this volume, for a critical discussion of the claim that neurophysiological distinctions will directly "trickle up" to psychology.)

From here, Kim says, three directions are open to the nonreductivist: one may become a Cartesian dualist (no option), one may become an epiphenomenalist (no option) or one may turn to reductionism in order to rescue the causal efficacy of the mental in a physically constituted world. What Kim proposes is a "conditional physical reductionism" (2005, p. 5): *if* mental phenomena enjoy causal efficacy, and most of us have strong intuitions that they do, they enjoy it in virtue of them being type identical to neural phenomena. Mental events can be causally efficacious only in virtue of them being reducible to neural events: "If mental phenomena are neural processes in the brain, there will be no special mystery about mental causation" (Kim, 2005, p. 153) and, therefore, "[r]eduction is the stopper that will plug the cosmic hole through which causal powers might drain away" (Kim, 2005, p. 68). Only reductionism will be able to vindicate mental causation in a way that is satisfying to the physicalist. (See Gillett, Chapter 4, this volume, for further development of Kim's ontological reductionism, which, however, leaves room for theory nonreductionism.)

But if we buy Kim's arguments for reductionism, what kind of reductionism can this be? In the absence of bridge laws, reductive explanation may still be possible (Kim, 2005, p. 97). Basically, reductive explanation, as Kim understands it, consists in a three-step procedure. The first step is functionalization: give a job description of the property that is to be reduced; specify its causal role. The second step is to find the physical realizer for the functionalized property. The third step is to provide an explanation of how the physical realizers fulfill the causal role specified in the first step. Examples are *gene* and *temperature* (see 1998, ch. 1). This is reduction without bridge laws: the relation between mental and neural is a role-filler relation. In accordance with the functionalizing strategy, Kim urges to look for local, presumably species-specific, reductions. Psychological states like representations or pain may have neurally different realizations in different organisms, and reducing them may produce a disjunctive series of local reductive identifications between mental and neural events. In this case, losing generalizations over the functional causes of the behavior of different species may even be good riddance.

4. New Chapters in Reduction: Metascience, Mechanicism, and Pluralism

4.1 Metascience and Mechanicism

In this section we will consider another alternative to the "sweeping," single-purpose accounts we considered above. Many have called attention to the fact that historical developments in science resist being captured in such uniform models. McCauley (Chapter 9, this volume), for instance, argues that reduction in science is neither simple nor unitary. The inadequacy of global accounts of reduction is particularly clear when one considers the life sciences. Here we don't typically find laws or sweeping, large-scale theories as was required by standard nomothetic accounts of explanation. In particular in biology and cognitive science, scientists' aims are at a much more local scale: they search for functionally characterized models of increasingly finer grain that explain selected phenomena at higher levels (Cummins, 2000). In such domains as the cognitive, biological, and neural sciences, researchers provide successful explanations without providing laws; they aim at uncovering and specifying mechanisms. Hence, as Richardson puts it (Chapter 6, this volume), what we see in science is not theory reduction, but a "succession of models constituting partial solutions based on inadequacies to specific and local problems," rather than the incorporation of one theory in the next one.

Theory reduction fares badly when one's goal is to describe what actually happens in science. Bickle argues that we should let go of the "philosopher's fantasies": classical reductionism, functional reductionism, and new wave reductionism (the latter developed by his own former self, among others). This is not to say that science is not reductionistic. The point is that we must get into the laboratory and look at actual science – from the bottom up, as it were – to find out what reduction is: reductionism can only be reductionism-in-practice (Bickle, 2003). We should leave behind philosophy and embrace ("new wave") metascience: clarifying reductionism is "letting a sense of reduction emerge from the detailed investigations drawn from recent scientific practice" (p. 31). Neuroscientific experimental and explanatory practices show that mind-to-molecule (or mind-to-cell) links are established all the time through what Bickle calls "intervene cellularly/molecularly and track behaviorally" approaches, i.e., lesioning, knocking out genes, or otherwise manipulating lower-level constituents of a system and then tracking the behavioral effects of such interventions. Molecular and cellular mechanisms are claimed to directly explain the behavioral data – and that's reduction if anything is! In his chapter Bickle argues that even consciousness, "the castle keep, the central redoubt, the core essence of true mentality" (Churchland, 1995, p. 212), might not be able to escape such "ruthless reductionism": science now offers clear views of the molecular mechanisms underlying certain aspects of consciousness at the macromolecular level of agonistic activities at subunits of γ-amino-butyric acid type A ($GABA_A$) receptor proteins.

Many of the authors in this volume share Bickle's naturalistic view that an understanding of reduction and reductive explanation should start in science. Bechtel, Clark, Bickle, Richardson, Wright, and McCauley (this volume) all point out that, especially in the life sciences, the search for and identification of mechanisms that are responsible for a phenomenon under investigation is of central importance. On "ruthless reductionism," however, we can and should descent immediately to the lowest possible levels, i.e., the levels of cells and molecules, and this allows us to "*set aside causal-mechanistic explanations* offered at intermediate levels of theorizing" (Bickle, Chapter 12, this volume). Bechtel, Clark, Richardson, Wright, and McCauley disagree. Clark, for instance, says that "intermediate-level analyses are of great importance." We should be careful to note, however, that claiming that higher-level analyses are important is not the same as claiming that there is no room for reductive explanation.

Bechtel claims that mechanistic explanations are reductionistic in their appeal to lower levels. In this respect, Clark speaks of "homucular explanations" which he sees as "the contemporary analogue to good old-fashioned *reductionistic* explanation." However, as both Bechtel and Clark tell us, looking down to lower-level – cellular, molecular, or systemic components, for instance – does not suffice. One must move beyond accounts of the parts of a mechanism and how they operate. The *organization* of the parts and interactions of the mechanism with its environment requires (semi-) autonomous higher-level research. Thus, some kind of autonomy for psychology can be maintained without multiple realizability since, according to Bechtel, higher-level accounts provide "additional information." Clark emphasizes that cognitive science should strive for "a satisfying and mutually illuminating interlock" between three different explanatory styles: homuncular, interactive, and emergent explanation. Homuncular explanation alone can never suffice when one's goal is to understand embodied, embedded agents. In such cases, entirely different explanatory strategies must be pursued. Thus, whereas Bickle takes molecular and cellular neuroscience to bypass higher-level analyses altogether, higher-level and lower-level investigations complement one another according to Bechtel, Clark, and others. Hence, these authors dismiss Bickle's explanatory monism according to which a behavioral phenomenon is explained by a single account furnished at the lowest possible level.

4.2 Pluralism and Co-evolution

The new mechanicists offer a very moderate kind of reductionism. They argue that it is important in the cognitive, biological, and neural sciences to specify lower-level mechanisms to explain selected higher-level phenomena, such as the behavior of systems under specified conditions. This does not involve the reduction or elimination of entire upper-level theories. The importance of higher-level theories, is not denied: explanatory ascent is as important as explanatory descent. Clark describes such an outlook as "explanatory liberalism." We will now see how these ideas concerning mechanistic analysis and emergent explanation can, according to some authors, be embedded in a more general view of explanatory pluralism.

Wimsatt (1976a) argued that not all reductions are of a uniform nature (as suggested by Nagel, Schaffner, and the new wave reductionists) and proposed to distinguish between two types of reduction, what he labeled "interlevel reduction" (roughly what Nickles (1973) had called "reduction$_1$") and "intralevel" or "successional reduction" (roughly Nickles's "reduction$_2$"). Intralevel reduction involves the relations between an older theory and a newer, succeeding theory (say, T_R and T_R^*), with the latter correcting the former. The intralevel or diachronous context is the context of intertheoretical relations that involves the modification and succession of theories over time. Such reductions concern transformational, possibly non-deductive and diachronous relations between theories (see also McCauley, 1986). Looking at successive scientific theories, one sees the *transformation* of theories in the light of mutual similarities and differences. Within-level reductions are about localizing, demonstrating and analyzing the analogies obtaining between theories T_R and theories T_R^*.

Interlevel (or explanatory) reductions, on the other hand, are of an altogether different kind. Wimsatt asserts that in contrast to the formal or structural models discussed above, we never find "total deductive systematization" as in classical models of reduction (like Nagel's or Schaffner's) and such global systematization is also "clearly unnecessary and irrelevant to the search for explanations" (Wimsatt, 1976b, p. 684). As Wimsatt pointed out many years before the current wave of mechanicism in the philosophy of science, biologists are reductionistic, not in the sense that they are interested in explaining theories through derivation, but because they aim at explaining phenomena by discovering mechanisms. Whereas for Nickles interlevel reductions are obtained by Nagel-style derivational reductions, according to Wimsatt (and this accords nicely with ideas formulated by mechanistic philosophers of science), interlevel contexts do not engage relations between theories at all, rather in such contexts one considers properties of higher-level entities and how they relate to properties of lower-level entities. What most scientists mean when they talk about reduction or reductive explanation is answering questions like: how is this or that phenomenon produced by causal interactions at lower levels? Here, Wimsatt explains, identificatory statements also play a role, but again not in the way envisaged by reductionists in the tradition of Smart, Sklar, Schaffner, and Kim: they are not ends in themselves, but rather tools that guide scientific progress. Scientists are not primarily interested in ontological claims of the sort $A = B$; rather their purposes are first and foremost of an explanatory nature. In Wimsatt's reconstruction of actual scientific practices, identity statements are hypothetical and heuristic and are used to detect and locate explanatory failures which in their turn drive intralevel theory changes (Wimsatt, 1976a, pp. 225–230). Wimsatt's suggestions thus embody a reading of the identity theory which was later developed in McCauley and Bechtel's heuristic identity theory (McCauley & Bechtel, 2001): "the optimal strategy for the identity theorist is not to waste time arguing for the in principle possibility of the identity theory, but to look for plausible explanations for the important and relevant differences between the mental and physical realms. If the explanations are forthcoming, the identities will be assumed. If not, the explanatory failures will force a careful use of Leibniz's Law to detect differences which might be used as the basis for new explanatory hypotheses" (p. 229).

Rather than theories being constantly under threat from lower-level ones, explanatory pluralists have discerned a "peaceful coexistence" between theories and models (McCauley, 1986; 1996; Schouten & Looren de Jong, 1999). They typically follow, again, Wimsatt's arguments developed in the 1970s: "Theoretical conceptions of entities at different levels co-evolve and are mutually elaborated . . . under the pressure of one another . . . [A]ll corrections in theory get packed into a 'successional' component and all unfalsified explanatory and compositional statements get packed into the 'explanatory reduction' component" (Wimsatt, 1976b, p. 682). Thus, distinct, though typically adjacent, levels mutually exert *selection pressures* and are engaged in a process of *co-evolution*. Bechtel, McCauley, and Wright show how this two-way flow of information works for the psychology-neuroscience, while Richardson speaks of "bidirectional exchange" between chemistry and physics.

Now we should note that it is certainly true that a number of philosophy's most uncompromising reductionists have recognized the importance of co-evolution (Bickle, 1998; Churchland, 1986). Psychology does have a role to play in developing explanations of behavior, even for reductionists. Hence, the Churchlands claim that "we count ourselves among the most fervent of the Friends of Psychology" (Churchland & Churchland, 1996, p. 219). However, one may doubt that this involves *genuine* co-evolution, with lasting contributions from both upper-level and lower-level theories (see Van Eck, Looren de Jong, & Schouten, 2006). For instance, Bickle's ideas on co-evolution amount to the view that psychological theories only provide fairly short-lived heuristics. After an *initial* co-evolutionary phase between theories at distinct levels, in which psychological theories provide crude descriptions of the phenomena to be explained, interlevel corrective, "structuring" influences between psychology and neuroscience travel from neuroscience to psychology, and not the other way around. The question for these reductionists is this: "Can we reconstruct all known *mental* phenomena in *neurodynamical* terms?" (Churchland, 1995, p. 211), a question which reductionists typically answer in the affirmative; neuroscientific results are simply fed into current psychology, which is then "simply *becoming* the Neuroscience of very Large and Intricate Brains" (Churchland & Churchland, 1996, p. 224). The inevitable outcome of this so-called co-evolutionary process will be that the neurosciences will be able to provide *exhaustive* fine-grained explanations, thereby rendering psychology explanatorily inert along the way. Bickle puts it thus: "There is no need to evoke psychological causal explanations, and in fact scientists stop evoking and developing them, *once real neurobiological* explanations are on offer" (Bickle, 2003, p. 110, original emphasis). Psychology, as Wright (this volume) puts it, simply becomes extinct.

In contrast, those with pluralist inclinations insist on *enduring* co-evolution, with higher-level sciences like psychology generating lasting influences on lower-level investigations. Wright, for instance, examines the mechanisms of motivation and brain reward function and shows that here preclusion of higher-level explanations would obstruct explanatory progress. He concludes that the idea of psychological explanations becoming extinct is a myth, not supported by scientific practice. Similar points are made by Endicott (Chapter 7, this volume) who argues that lower-level explanations require reference to higher-level properties. Reductionism fails because it does not do justice to the role of higher-level theories. Actual science shows that resources drawn from higher levels continue to play a role. In all, explanatory pluralism offers a view of scientific progress that highlights the fact that science works in a local, piecemeal fashion.

4.3 Pluralism and the Metaphysics of Science

The brief review of recent developments above indicates that reduction is a far more complex and dynamical affair than the classical picture suggested. The arrow of reduction is complemented with higher-level constraints downward; reduction can go hand in glove with higher-level explanations. Thus, reduction and autonomy are not necessarily contradictory (As the title of Bechtel's Chapter 8 in this volume shows). Looking back, we can now see that [CC] and [DC] were a concern for philosophers, generated by the Logical Positivist view of theories, not a real problem in science. The failure of finding bridge laws between two sets of theories and the unruly relations between them are part of the ongoing dynamics of scientific progress. Interestingly, the same rejection of the reduction/autonomy dichotomy can be seen in the metaphysical, if you will, metatheoretical contributions in this volume. In different ways they show the compatibility of reduction with a legitimate role for higher-level explanations. Reduction vs. autonomy is a "false dichotomy," according to Gillett, a view shared by Melnyk, Shapiro, and Polger: reductionism and antireductionism offer a "false choice" (Polger); functionalism and reductionism are "friends not foes" (Melnyk), and current empirical developments suggest "reduction *of a sort*" and "autonomy *of a sort*" (Shapiro).

Gillett presents a metaphysics of science, in particular analyses of the nature of the compositional relations, to underpin the mechanistic explanations (as provided by authors like Bechtel and Richardson) involved in such explanations. These compositional relations, uncovered in scientific investigations, drive a form of metaphysical reductionism: since the composing entities non-causally determine higher-level (composed) entities, the latter must be illusory. Nevertheless, it is argued, "New" reductionism is compatible with the nonreductivist's claim that the predicates, concepts and theories of the special higher-level sciences are (in principle) indispensable.

Melnyk too sees reductionism, properly understood, as compatible with functionalism, once the mainstay of antireductionism and autonomy. Psychological phenomena are multiple realizable, psychological explanations pick out really existing patterns, and psychological explanations may be used to revise explanations in terms of physical phenomena – as pluralists would agree. In fact, the possibility of mutual co-evolutionary feedback between psychology and neuroscience even presupposes some form of metaphysical reductionism, argues Melnyk.

Shapiro argues against Kim's claim that multiple realization leads to local reductions, and thus to the disintegration of psychology. Shapiro points out that not all causal differences in the lower-level physical properties are relevant for the higher mental level. And since these differences do not necessarily "trickle up," it is not obvious that psychology will fractionate along the lines of physical kinds. However, he also argues that psychology now faces disintegration from another direction, viz., theories of embodied cognition. These may, in Shapiro's view, have the consequence that there are as many subdisciplines of psychology as there are types of body.

Polger distinguishes various approaches to reductionism and argues that they all result from the problematic assumption that there is only one ontology and one true story of the world. He argues that there is more than one ontology and more than one explanation for a phenomenon. Hence, he defends an approach which, he says, is "genuinely nonreductive" in the sense that it is neither reductive nor antireductive, but pluralistic (or naturalistic).

To sum up, the dichotomy between reductionism and autonomy that we started with is a simplification. Careful conceptual work in the metaphysics of science (Gillett, Polger, Melnyk, Shapiro), in empirically informed work in the philosophy of science (Clark, Richardson, Endicott, Bechtel, McCauley), and empirical case studies and laboratory work in neuroscience (Wright, Bickle, Looren de Jong & Schouten) yield the picture of many connections and, in Kant's terminology, of passages between the many levels and domains of study of the mind/brain. The most reductionist position is defended by John Bickle, whose strategy of confronting philosophical problems (and philosophers) with the latest data from the laboratory bench is exceptional, but yields stimulating results. Most of the other authors, however, will acknowledge that to a more or lesser degree higher-level explanations are indispensable, but not autonomous; and that psychology and neuroscience are and should be connected and perhaps integrated, but not unified along physicalist lines.

5. Gaps and Gulfs: Unity and Pluralism

Recall that in this overview of the territory we departed on the assumption that unity is a crucial desideratum in science. As Klein and Lachièze-Rey say: "without unity as a beacon, the world, indeed human thought itself, would scatter into a dust of things and ideas impossible to integrate" (1999, p. vii). Such a unity of science apparently involves what Descartes once called a *catena scientiarum* (*Cogitationes privatae*, AT 10: 215). Scientific disciplines and theories must be strung together. And science is, some argue, well on its way toward such a concatenated unity: the physicist Steven Weinberg once remarked that in science we can see "a convergence of the arrows of explanation, like the convergence of meridians towards the North Pole. Our deepest principles, although not yet final, have become steadily more simple and economical" (Weinberg, 1993, pp. 231–232). Here we see a succinct formulation of reductionism's fundamental aims: nothing less than a "final" and "simple and economical," unified view of reality. Thus, reductionism is often taken to be committed to an explanatory monism which is supposed to deliver something high on science's wish list: a unity of knowledge (see, however, the chapters by Melnyk, Polger, and Gillett in this volume). We have also seen however that it turned out to be difficult, if not impossible, to formulate what it means to reductively concatenate theories in a way that does justice to living science. Thus, our overview of the arguments pro and contra reductionism led to explanatory pluralism instead of explanatory monism.

So, one might ask, haven't explanatory pluralists sacrificed our cherished ideals of unity and integration? Not necessarily. One option that might be explored is to say "so much the worse for the unity of science" (e.g., Cartwright, 1999; Dupré, 1983; Fodor, 1981; Van der Steen, 1993), or one might argue that is was mistaken to tie unification to reductionism in the first place. For reasons of space we shall not go into the first option. It is the last option (of integration-without-reduction) that we will explore a little further in the remainder of this introductory chapter.

It is well known that the very idea of an *Einheitswissenschaft* was made famous by logical empiricism. However, we must be careful to note that opinions within the broader logical-empiricist movement strongly diverged on this issue. Not everyone in the movement agreed with Nagel that reductionism offered the royal road to a unified science.

Nagel's ideas concerning a reductionist construal of unification were foreshadowed in a paper ("The logic of reduction in the sciences") read at the movement's Prague conference of 1934 and later published, along with the other papers presented at this conference, in the movement's house organ *Erkenntnis.* Read at the same conference (and published in the same volume of *Erkenntnis*) was a paper by Otto Neurath, entitled "Unity of science as a task" (Neurath, 1935). Here Neurath emphasizes that the unity of science does not involve a unity of laws – which, for instance, Carnap (see, e.g., 1938) wished to defend, but a mere unity of language. Although clearly a unificationist, Neurath was not a reductionist. Whereas Carnap thought it possible to ultimately derive the sociological laws from the laws of physics, Neurath dismissed this possibility: "*The development of physicalist ontology does not mean the transfer of laws of physics to living things and their groups*, as some have thought possible." There is no need to "go back to the microstructure, and thereby to build up these sociological laws from physical ones" (Neurath, 1931, p. 75). Physicalism in Neurath's sense only requires that the sociologist (or psychologist) speaks of entities observable in space and time and describable in what he called the "Universal Jargon."

Neurath markets the idea of a unified science (of which he was the spiritual father) as "encyclopedic integration" and this did not involve anything like looking for a "super-science" (Neurath, 1937, p. 265). As Neurath remarks, "'*The system*' *is the great scientific lie*" (Neurath, 1935, p. 116), because, and here his statements are a distant echo of the Heraclitean *panta rhei*, "basically everything is fluid, . . . multiplicity and uncertainty exist in all science. . . . *The whole of science is basically* always under discussion" (p. 118). "*Alles fließt*" in science (cited in Reisch, 1998), and this observation indicates that Neurath was not a reductionist (see also Uebel, 2000) and a pluralist. Unified science has everything to do with a "pluralist attitude" as there is no comprehensive worldview and the encyclopedia remains full of "gaps and gulfs" (Neurath, 1946, p. 497). Encyclopedism is all about antitotalitarianism, tolerance, and *laissez faire.* In a spirit very close to what is upheld by the explanatory pluralists, Neurath adds that

> our scientific practice is based on local systematizations only, not on overstraining the bow of deduction. Very often scientists know perfectly well that certain principles applied to a certain area are very fruitful, while contradictory principles applied to a different area also appear to be fruitful. It would, of course, be nice to harmonize the demonstrations in both areas, but in the meantime, scientific research progresses successfully. (p. 498)

The pluralism involved in unified science can, according to Neurath, be best understood through Horace M. Kallen's (1946) metaphor of the *orchestration* which involves "diversities of instruments and parts, of movements and pauses, of dissonances and discords as well as harmonies" (Kallen, 1946, pp. 495–496).

So, going back to the roots of empiricism suggests a more pluralist view than the ruling consensus established by later generations. As mentioned, pluralism emerges in this volume both from conceptual and metaphysical analyses on the one hand, and from case studies on the other hand. To sum up, connecting domains of knowledge is not necessarily bringing higher levels under the rule of physics (if there is such a rule). As Kant put it (*Kritik der Urteilskraft*, 1799, p. 305), each science is a separate, whole structure, although a connection (passage, *Übergang*) can be made afterwards between them. Kant's metaphor may be too static (in Kant's own words, architectonic) for our taste, but we would like to adopt the image of passages. Connecting scientific domains may not at all be like annexating and rebuilding psychology by neuroscience, as reductionists suggest, but more like building passages between one part of the many semi-detached buildings of science and another.

Acknowledgments

We thank Cory Wright and Dingmar van Eck for many useful criticisms and suggestions.

Note

1 We owe this example to Cory Wright.

References

Bechtel, W., & Mundale, J. (1999). Multiple realizability revisited: Linking cognitive and neural states. *Philosophy of Science*, 66, 175–207.

Bickle, J. (1998). *Psychoneural Reduction: The New Wave*. Cambridge, MA: MIT Press.

Bickle, J. (2003). *Philosophy and Neuroscience: A Ruthlessly Reductive Account.* Dordrecht: Kluwer Academic Publishers.

Brandt, R., & Kim, J. (1967). The logic of the identity theory. *Journal of Philosophy*, 64, 515–537.

Brittan, G. G. (1970). Explanation and reduction. *Journal of Philosophy*, *67*, 446–457.

Brooks, D. H. M. (1994). How to perform a reduction. *Philosophy and Phenomenological Research*, *54*, 803–814.

Caplan, A. (1981). Babies, bathwater and derivational reduction. *PSA 1978*, *2*, 357–370.

Carnap, R. (1938). Logical foundations of the unity of science. In *International Encyclopedia of Unified Science* (Vol. I, No. 1, pp. 42–62). Chicago: University of Chicago Press.

Cartwright, N. (1999). *The Dappled World: A Study of the Boundaries of Science.* Cambridge: Cambridge University Press.

Causey, R. L. (1972). Attribute-identities in microreductions. *Journal of Philosophy*, *79*, 407–422.

Churchland, P. M. (1995). *The Engine of Reason, the Seat of the Soul: A Philosophical Journey into the Brain.* Cambridge, MA: MIT Press.

Churchland, P. M. (2005). Functionalism at forty: A critical retrospective. *Journal of Philosophy*, *102*, 33–50.

Churchland, P. M., & Churchland, P. S. (1996). Replies, A: The future of psychology, folk and scientific. In R. N. McCauley (Ed.), *The Churchlands and Their Critics* (pp. 219–255). Cambridge, MA: Blackwell.

Churchland, P. S. (1986). *Neurophilosophy: Toward a Unified Science of the Mind–Brain.* Cambridge, MA: MIT Press.

Cummins, R. (2000). "How does it work" versus "What are the laws": Two conceptions of psychological explanation. In F. C. Keil & R. A. Wilson (Eds.), *Explanations and Cognition* (pp. 117–144). Cambridge, MA: MIT Press.

Dawkins, R. (1982). *The Extended Phenotype: The Long Reach of the Gene.* Oxford: Oxford University Press.

Dennett, D. C. (1995). *Darwin's Dangerous Idea: Evolution and the Meanings of Life.* New York: Simon & Schuster.

Dupré, J. (1983). The disunity of science. *Mind*, *92*, 321–346.

Enç, B. (1976). Identity statements and microreductions. *Journal of Philosophy*, *63*, 285–306.

Enç, B. (1983). In defense of the identity theory. *Journal of Philosophy*, *80*, 279–298.

Feigl, H. (1958). The "mental" and the "physical." In H. Feigl, M. Scriven & G. Maxwell (Eds.), *Concepts, Theories, and the Mind–Body Problem: Minnesota Studies in the Philosophy of Science* (Vol. 2, pp. 370–497). Minneapolis: University of Minnesota Press.

Feyerabend, P. K. (1965). Reply to criticism. In R. S. Cohen & M. W. Wartowsky (Eds.), *Boston Studies in the Philosophy of Science. Vol. II: In Honour of Philipp Frank* (pp. 223–261). New York: Humanities Press.

Feyerabend, P. K. (1981). Explanation, reduction and empiricism. In P. K. Feyerabend (Ed.), *Realism, Rationalism and Scientific Method. Philosophical Papers* (pp. 44–96). Cambridge, MA: Cambridge University Press.

Fodor, J. (1974). Special sciences, or the disunity of science as a working hypothesis. *Synthese*, *28*, 97–115. Reprinted in N. Block (Ed.), *Readings in Philosophy of Psychology* (Vol. 1, pp. 120–133). Cambridge, MA: Harvard University Press, 1980.

Fodor, J. (1998). *In Critical Condition: Polemical Essays on Cognitive Science and the Philosophy of Mind.* Cambridge, MA: MIT Press.

Fodor, J. A. (1975). *The Language of Thought.* New York: Thomas Y. Crowell.

Fodor, J. A. (1981). Special sciences, or the disunity of science as a working hypothesis. In J. A. Fodor, *RePresentations: Philosophical Essays on the Foundations of Cognitive Science* (pp. 127–145). Brighton, UK: Harvester Press.

Gaa, J. (1975). The replacement of scientific theories: Reduction and explication. *Philosophy of Science*, *42*, 349–372.

Hempel, C. G. (1969). Reduction: Ontological and linguistic facets. In S. Morgenbesser, P. Suppes & M. White (Eds.), *Philosophy, Science, and Method: Essays in Honor of Ernest Nagel* (pp. 179–199). New York: St Martin's Press.

Hooker, C. A. (1981). Towards a general theory of reduction. *Dialogue*, *20*, 38–59, 201–236, 496–529.

Hull, D. L. (1974). *Philosophy of Biological Science.* Englewood Cliffs, NJ: Prentice-Hall.

Kallen, H. M. (1946). The meanings of "unity" among the sciences, once more. *Philosophy and Phenomenological Research*, *6*, 493–496.

Kemeny, J. G., & Oppenheim, P. (1956). On reduction. *Philosophical Studies*, *7*, 6–19.

Kim, J. (1993). *Supervenience and Mind: Selected Philosophical Essays.* Cambridge: Cambridge University Press.

Kim, J. (1998). *Mind in a Physical World: An Essay on the Mind–Body Problem and Mental Causation.* Cambridge, MA: MIT Press.

Kim, J. (2005). *Physicalism, or Something Near Enough.* Princeton: Princeton University Press.

Klein, E., & Lachièze-Rey, M. (1999). *The Quest for Unity: The Adventure of Physics.* Oxford: Oxford University Press.

McCauley, R. N. (1986). Intertheoretic relations and the future of psychology. *Philosophy of Science*, *53*, 179–199.

McCauley, R. N. (1996). Explanatory pluralism and the co-evolution of theories in science. In R. N. McCauley (Ed.), *The Churchlands and Their Critics* (pp. 17–47). Cambridge, MA: Blackwell.

McCauley, R. N., & Bechtel, W. (2001). Explanatory pluralism and heuristic identity theory. *Theory and Psychology*, *11*, 736–760.

McRae, R. (1957). Kant's conception of the unity of the sciences. *Philosophy and Phenomenological Research*, *18*, 1–17.

Nagel, E. (1961). *The Structure of Science: Problems in the Logic of Scientific Explanation*. London: Routledge & Kegan Paul.

Nagel, E. (1970). Issues in the logic of reductive explanations. In H. E. Kiefer & M. K. Munitz (Eds.), *Mind, Science, and History* (pp. 117–137). Albany, NY: State University of New York Press.

Neurath, O. (1931). Sociology in the framework of physicalism. In R. S. Cohen & M. Neurath (Eds.), *Philosophical Papers 1913–1946* (pp. 58–90). Dordrecht: Kluwer Academic Publishers.

Neurath, O. (1935). The unity of science as a task. In R. S. Cohen & M. Neurath (Eds.), *Philosophical Papers 1913–1946* (pp. 115–120). Dordrecht: Kluwer Academic Publishers.

Neurath, O. (1937). Unified science and its encyclopedia. *Philosophy of Science*, *4*, 265–277.

Neurath, O. (1946). The orchestration of the sciences by the encyclopedism of logical empiricism. *Philosophy and Phenomenological Research*, *6*, 496–508.

Nickles, T. (1973). Two concepts of intertheoretic reduction. *Journal of Philosophy*, *70*, 181–201.

Place, U. T. (1956). Is consciousness a brain process? *British Journal of Psychology*, *47*, 44–50.

Polger, T. W. (2004). *Natural Minds*. Cambridge, MA: MIT Press.

Popper, K. R. (1957). The aim of science. *Ratio*, *1*, 24–35.

Putnam, H. (1960). Minds and machines. In S. Hook (Ed.), *Dimensions of Mind: A Symposium* (pp. 148–179). New York: New York University Press.

Putnam, H. (1965). How not to talk about meaning. In R. S. Cohen & M. W. Wartowsky (Eds.), *Boston Studies in the Philosophy of Science. Vol. II: In Honour of Philipp Frank* (pp. 205–222). New York: Humanities Press.

Putnam, H. (1975). Philosophy and our mental life. In H. Putnam, *Mind, Language and Reality: Philosophical Papers* (pp. 291–303). Cambridge, MA: Cambridge University Press.

Reisch, G. A. (1998). Pluralism, logical empiricism, and the problem of pseudoscience. *Philosophy of Science*, *65*, 333–348.

Schaffner, K. F. (1967). Approaches to reduction. *Philosophy of Science*, *34*, 137–147.

Schouten, M. K. D., & Looren de Jong, H. (1999). Reduction, elimination, and levels: The case of the LTP-learning link. *Philosophical Psychology*, *12*, 237–262.

Sellars, W. (1965). Scientific realism or irenic instrumentalism. In R. S. Cohen & M. W. Wartofsky (Eds.), *Boston Studies in the Philosophy of Science. Volume 2: In Honor of Philipp Frank* (pp. 171–204). New York: Humanities Press.

Shapiro, L. A. (2004). *The Mind Incarnate.* Cambridge, MA: MIT Press.

Sklar, L. (1967). Types of inter-theoretic reduction. *British Journal for the Philosophy of Science*, *18*, 109–124.

Smart, J. J. C. (1959). Sensations and brain processes. *Philosophical Review*, *68*, 141–156.

Suppes, P. (1981). The plurality of science. *PSA 1978*, *2*, 3–16.

Uebel, T. (2000). *Vernunftkritik und Wissenschaft: Otto Neurath un der erste Wiener Kreis.* Vienna and New York: Springer.

van der Steen, W. J. (1993). Towards disciplinary disintegration in biology. *Biology and Philosophy*, *8*, 259–275.

van Eck, D., Looren de Jong, H., & Schouten, M. K. D. (2006). Evaluating new wave reductionism: The case of vision. *British Journal for the Philosophy of Science*, *57*, 167–196.

Weinberg, S. (1993). *Dreams of a Final Theory.* New York: Pantheon Books.

Wilson, E. O. (1998). *Consilience: The Unity of Knowledge.* New York: Alfred A. Knopf.

Wimsatt, W. C. (1976a). Reductionism, levels of organization, and the mind–body problem. In G. G. Globus, G. Maxwell & I. Savodnik (Eds.), *Consciousness and the Brain: A Scientific and Philosophical Inquiry* (pp. 205–267). New York: Plenum.

Wimsatt, W. C. (1976b). Reductive explanation: A functional account. *PSA 1974*, 671–710.

PART I

METAPHYSICS OF SCIENCE

2

FUNCTIONALISM AND PSYCHOLOGICAL REDUCTIONISM: FRIENDS, NOT FOES

Andrew Melnyk

1. Introduction

Some opponents of psychological reductionism oppose it because they reject physicalism about the mind; they hold that psychological properties, and perhaps even minds themselves, are at most causally or nomologically dependent on the physical properties of brains and their physical environments.[1] However, there are other opponents of psychological reductionism who accept physicalism, understanding minds and their properties to be, in some suitably broad sense, physical. It is to these physicalist opponents of psychological reductionism that I'll speak in this chapter.

Here are four claims that such opponents typically make by way of complaint about psychological reductionism:

(1) Psychological phenomena are often multiply realized.

(2) The explanations that psychologists supply are more than merely heuristic: they identify objectively existing explanatory factors that the explanations of neurophysiologists (or molecular biologists, or whoever) fail to identify.

(3) It is possible for psychologists to discover psychological entities, psychological processes, and the psychological generalizations that

govern them without either investigating or already knowing about the underlying physical (e.g., neurophysiological or biochemical) phenomena.

(4) It is possible for discoveries in psychology to *correct* claims made by those who study the underlying physical phenomena.

Now such claims as these are plausible in their own right. But they have often been seen also – and sensibly – as natural consequences of a *broadly functionalist* picture of psychological phenomena.[2] According to this picture, the physical entities of which nature is ultimately composed can come, in accordance with physical laws, to exhibit certain fully objective patterns of organization. But although these patterns are patterns *in* physical entities, and in nothing else, they are still independent of physical entities in the sense that the very same patterns can be exhibited in systems with greatly varying physical compositions (claim 1). Consequently, these patterns can be described and explained in their own right, and it is precisely the role of the special sciences, including psychology, to do so (claim 2). The entities studied by psychology, then, are *functional* (though physically realized) entities: they are entities whose existence just *is* the existence of some or other (physical) entity or complex of (physical) entities that exhibits such-and-such a pattern, where patterns can be specified by reference to causal or computational roles, or indeed in other ways. Thus, psychologists who investigate these entities do not necessarily require knowledge framed using the conceptual repertoires of physical sciences such as neurophysiology or biochemistry (claim 3). However, precisely because psychological phenomena are patterns in *physical* phenomena, the discovery of psychological phenomena that *couldn't* be realized in physical phenomena as we currently conceive them would be evidence that our current conception of physical phenomena is incorrect (claim 4).[3]

Since these four claims issue from a broadly functionalist picture of psychological phenomena, it's tempting to think that such a picture is *inconsistent* with psychological reductionism; and many philosophers have yielded, and continue to yield, to this temptation (e.g., classically, Fodor, 1974 and Putnam, 1975, ch. 14; and, recently, Ross & Spurrett, 2004). But, I claim, they've been wrong to do so, for, as I'll argue in this chapter, a broadly functionalist picture of psychological phenomena is quite consistent with at least one interesting thesis of psychological reductionism. I'll work toward a formulation of this thesis of psychological reductionism in Section 2. Then, in Section 3, I'll argue that psychological reductionism, thus formulated, is consistent with a broadly functionalist picture of psychological phenomena by showing how it is consistent with the four supposedly antireductionist claims stated above.[4]

2. Reducibility and Psychological Reductionism

In this section, I'll first describe the kind of reducibility presupposed by the interesting thesis of psychological reductionism mentioned above. (There may well be other kinds of reducibility, but if there are, I won't be talking about them.) Then I'll formulate the thesis.

2.1 Reducibility

The first thing to say about the kind of reducibility presupposed by the thesis of psychological reductionism that we'll eventually formulate is that it's a relation between contingent phenomena of one kind (e.g., biological phenomena) and contingent phenomena of another kind (e.g., chemical phenomena). The phenomena fall into two categories: first, particular phenomena, such as the existence of object-tokens, property-instances, and process-tokens; second, general phenomena, such as the holding of *ceteris paribus* regularities. (So my use of "phenomena" is very liberal.) What sort of relation is reducibility? It's best viewed, I suggest, as a special kind of non-causal, synchronic explainability. So, for example, biological phenomena are reducible to chemical phenomena if biological phenomena can be non-causally, synchronically explained – in a special way – by chemical phenomena, including chemical regularities. For the chemical phenomena to explain the biological phenomena in my intended sense of "synchronically," some (but not necessarily all) of the chemical phenomena must be simultaneous with the biological phenomena they explain.

For biological phenomena to be non-causally, synchronically explained by chemical phenomena in the *special way* mentioned above, two conditions must be met. The first condition is that purely chemical phenomena must be the *only contingent* phenomena involved in the explaining. Therefore, we *wouldn't* have reducibility if the non-causal, synchronic explanation of biological phenomena by chemical phenomena required the holding of a fundamental chemical-to-biological law of emergence. Because the holding of such a law, though a contingent phenomenon, wouldn't be a *purely* chemical phenomenon (since it would be partly biological), we wouldn't have the reducibility of biological to chemical phenomena.

The second condition for explanation in the special way mentioned above is that the metaphysical necessitation of biological phenomena by the chemical phenomena that explain them must be the metaphysical necessity of relevant object- or property-identities, e.g., identities between biological properties and chemical properties, or between biological properties and chemically realized functional properties. Therefore, if biological phenomena are reducible to chemical phenomena, the metaphysical necessitation of the biological phenomena by the chemical phenomena that explain them can't be any kind of *brute* (i.e., inexplicable) metaphysical necessitation between chemical phenomena and *entirely distinct* biological phenomena. Such brute metaphysical necessitation between distinct existences may in the end make no sense, as Hume famously thought; but if it does make sense, then it cannot be part of a genuinely *reductive* explanation. The reason is that if we learn that phenomena of one kind are reducible to phenomena of another kind, then we have discovered that we can shrink our overall ontological commitments; we no longer need to believe in the reducible phenomena as something *over and above* the reducing phenomena, and we no longer need to treat laws that link the reducing phenomena to the reducible phenomena as fundamental laws of nature. But if the allegedly reducing phenomena brute-metaphysically necessitate the allegedly reducible phenomena, then the latter phenomena are *still* something over and above the allegedly reducing phenomena, and so we have no genuine reducibility.

2.2 Psychological Reductionism

Any thesis of reductionism that invokes the kind of reducibility just described will obviously assert that phenomena of one kind are reducible to phenomena of another kind. But what are the two kinds of phenomena in the case of *psychological* reductionism?

Let us ask first which phenomena psychological reductionism should assert to be reducible. The obvious answer is "Psychological phenomena – psychological objects, properties, processes, regularities, and so on," but we need to say more precisely what these are. A natural first thought is to tie psychological phenomena to current psychology, and to take "psychological phenomena" to refer to all those phenomena asserted to exist, or of the *kinds* asserted to exist, in the consensus theories of current psychology.[5] But this proposal suffers from two problems. First, since the consensus theories of current psychology are bound to contain errors in what they assert to exist, some of the phenomena apparently picked out by the proposal will turn out not to exist. But only real phenomena are reducible in the present sense, since reducibility is a kind of explainability, and only genuine phenomena can be explained. Hence, if psychological reductionism claims the reducibility of all psychological phenomena, then on the present proposal it will be committed – absurdly – to the reducibility of certain non-entities. Second, since the consensus theories of current psychology are bound to be incomplete, in the sense of failing to mention some phenomena truly asserted to exist, or of kinds truly asserted to exist, in *future* psychology, a thesis of psychological reductionism that merely asserts the reducibility of all the phenomena asserted to exist, or of the kinds asserted to exist, in the consensus theories of *current* psychology will fail to assert the reducibility of *all* psychological phenomena.

Let us therefore take "psychological phenomena" to refer to all the phenomena *truly* asserted to exist, or of the kinds *truly* asserted to exist, in both the consensus theories of current psychology *and* the consensus theories of psychology at each point in its future. The requirement that psychological phenomena be those *truly* asserted to exist in certain theories avoids the problem resulting from the likely falsehood of some psychological theories, present *and* future. And the reference to *future* psychological theories avoids the problem of failing to include within the scope of psychological reductionism all the phenomena that it should include. It might be objected to this proposal that we have no principled way of telling whether a future science *is* psychology, given that we are assuming that psychology might change the content of its theories over time. But there is a way of tracking psychology over time that doesn't rely on the content of its theories: we can count any future science as psychology if it seeks, at least in the first instance, to explain *human behavior* – and perhaps also to give an account of the nature of such paradigmatic mental phenomena as beliefs, desires, and sensations.

Let us now ask to which phenomena psychological reductionism should claim that psychological phenomena are reducible. A tempting first response is to suggest that psychological reductionism should claim that psychological phenomena are reducible to *neurophysiological* phenomena that exist entirely within the skin of the thinker. But this response is inadvisable at least for the following reason.[6] If psychological reductionism asserts the reducibility of psychological phenomena to neurophysiological phenomena of this sort, then it renders itself incompatible with at least two kinds of plausible psychological hypothesis. The first kind of hypothesis identifies the representational content of an organism's mental states with those states' standing in certain relations to phenomena that are *external* to the organism (see, e.g., Fodor, 1990). If a hypothesis of this first (i.e., semantically externalist) kind is true, then not all psychological phenomena are reducible to neurophysiological phenomena. The second kind of plausible psychological hypothesis aims to "biologize" psychological phenomena by treating them as essentially possessing certain biological functions (see, e.g., Millikan, 1984; Lycan, 1987, ch. 4; Dretske, 1988). But the possession by an organism of a biological function is plausibly thought to require the organism to have had a certain evolutionary history in a certain environment. If so, then the truth of a hypothesis of this second kind would also entail that psychological phenomena are not reducible to (narrow) neurophysiological phenomena.

A better response to our question, then, is to say that psychological reductionism should claim that psychological phenomena are reducible to *non-psychological* phenomena. Non-psychological phenomena would obviously include (narrow) neurophysiological phenomena, but would also include phenomena in the *environment*, or in the *history*, of the organism that exhibits the psychological phenomena. This suggestion thus makes room for the reducibility of psychological phenomena that don't supervene solely on intrinsic and simultaneous features of the organism exhibiting the psychological phenomena.[7] However, we must construe "non-psychological phenomena" carefully. We can't take the term to refer to all phenomena not identical with psychological phenomena, since in that case psychological reductionism would have to be deemed *false* if psychological phenomena turned out to be *identical* with (say) neurophysiological phenomena! So we should take "non-psychological phenomena" to refer to those phenomena truly asserted to exist, or of the kinds truly asserted to exist, in the consensus theories of sciences other than psychology, both current and future. And, for this purpose, we should include our common-sense view of the physical world as an honorary science, since a reductive explanation of psychological phenomena might well have to invoke everyday objects such as cars and human bodies.

Let me now pull the various threads together. The interesting thesis of psychological reductionism that was promised in the introduction can be formulated like this:

> [Psychological Reductionism] All those phenomena truly asserted to exist, or of the kinds truly asserted to exist, in both the consensus theories of current psychology and the consensus theories of psychology at each point in its future are synchronically explainable by contingent phenomena that include only those phenomena truly asserted to exist, or of the kinds truly asserted to exist, in the consensus theories of sciences other than psychology, both current and future; but such explainability may not depend on any brute metaphysical necessitation of the psychological phenomena by entirely distinct phenomena.

All uses of "psychological reductionism" in the rest of this chapter will refer to this thesis.

3. The Consistency of Psychological Reductionism with the Functionalist Picture

Let me now argue that psychological reductionism is consistent with a broadly functionalist picture of psychological phenomena by showing that it is consistent with the four supposedly antireductionist claims that issue from the broadly functionalist picture. I'll consider each claim in turn.

3.1 The Multiple Realization of Psychological Phenomena

The first supposedly antireductionist claim was that psychological phenomena are often multiply realized. However, psychological reductionism is consistent with this claim because the multiple realization of psychological phenomena does not prevent their possessing an explanation, in terms of non-psychological phenomena, of the special sort required by psychological reductionism. To show that this is so, I will outline schematic representations of reductive explanations of multiply realized psychological phenomena. As already noted, psychological phenomena fall into the two categories of particular phenomena, such as psychological object-tokens and property-instances, and general phenomena, such as the holding of ceteris paribus psychological regularities. I will take each category of psychological phenomena separately, starting with particular phenomena.

Suppose, then, that a psychological state-type, P, is multiply realized by non-psychological phenomena. (Roughly, what this amounts to is, first, that P is a functional state-type of a particular sort, the state-type (say) that is tokened if some or other state-type is tokened that meets a certain specific condition, C; and, second, that, on at least two occasions on which P is tokened, it is *distinct* non-psychological state-types that, on each occasion, meet condition C.) Suppose, further, that P is tokened on a particular occasion. Here is how, in highly idealized form, we could represent a reductive explanation of this particular psychological phenomenon, even though P is a multiply realized psychological state-type:

(1) There is a tokening of a non-psychological state-type, N.

(2) There hold certain non-psychological regularities and (perhaps) there are tokenings of non-psychological types other than N.

So (3) there is a tokening of a state-type that meets condition C.

(4) There being a tokening of some or other state-type that meets condition C just *is* there being a tokening of P.

So (5) there is a tokening of P.

Premises (1) and (2) together describe the non-psychological phenomena to which the tokening of P is reducible.[8] Premise (1) describes the local non-psychological reduction base of the tokening of P, while premise (2) describes the existence of whatever non-local, non-psychological environmental or historical phenomena are required, and the holding of whatever non-psychological regularities are required, for the tokening of N to meet condition C and hence to realize a tokening of P. Therefore, (3) follows from (1) and (2). Premise (4) affirms the identity of the psychological state-type P with a certain functional state-type. From (4) and (3) there follows the conclusion (5), which reports the tokening of P that is to be reductively explained. I first presented a reductive derivation of this kind in Melnyk 1995, developing the idea in Melnyk 2003 (ch. 3), where it is defended much more fully. Jaegwon Kim has also defended a very similar idea (most recently in 2005, ch. 4). However, there are three important differences between Kim's position and mine. First, I insist that the identity of reduced types with functional types will almost always be an a posteriori matter, while Kim treats such identities as knowable a priori, presumably through a priori conceptual analysis. Second, Kim holds that reductive explanations somehow license the *identification* of instances of reduced properties with instances of reducing properties; I don't share this view. Third, Kim doesn't discuss how to reduce *regularities*, whereas I have done so in the sources cited, and will do so again below.

Now, the explanation that this derivation represents is indeed of the special kind described in my earlier account of reducibility, and hence qualifies as reductive.[9] First, it appeals in part to a particular non-psychological phenomenon (i.e., the tokening of N) that is simultaneous with the particular psychological phenomenon being explained. Second, the only contingent explanatory factors that the explanation appeals to are non-psychological phenomena. It is true that in the *representation* of the explanation premise (4) is required to serve as a bridge law to connect the disparate vocabularies in which representations of non-psychological and psychological phenomena are couched; but because premise (4), though not a priori, still expresses a necessary truth, it cites no explanatory factor in addition to those cited in premises (1) and (2).[10] Third, the explanation does not require the holding of a relation of brute metaphysical necessitation between distinct existences (i.e., between non-psychological and psychological phenomena). It's true that premise (4) expresses a necessary identity, as already noted, but identity is a relation between an entity and itself, not between an entity and some distinct entity.

Furthermore, and crucially, the reductive explanation represented by the derivation above is fully consistent with the multiple realizability of psychological state-type P. For, although the reductive explanation requires the identity of P with a *functional* state-type, it does not require the identity of P with any specific neurophysiological or biochemical or physical state-type. So it leaves open the possibility that some non-psychological state-type N*, distinct from N, should, given the right non-psychological environment and regularities, also meet condition C and hence that a token of N* should constitute the local reduction base of a distinct tokening of P – which would mean, of course, that P was multiply realized.

Let me now show that the reducibility of *general* psychological phenomena (i.e., of psychological regularities) is likewise consistent with the multiple realizability of psychological kinds.

Suppose that it is a psychological regularity (perhaps a *ceteris paribus* law) that P-tokens are followed by P*-tokens. Then, if P and P* are multiply realized psychological types, different *instances* of the regularity (i.e., different sequences each consisting of a particular P-token followed by a particular P*-token) will be differently realized by non-psychological phenomena. How, then, can the regularity have a reductive explanation? It can do so, I say, if *every instance* of the regularity has a reductive explanation; no more is required.[11] There is no conflict with multiple realizability, because the respective reductive explanations of the various instances of the regularity need not each appeal to the *same kind* of non-psychological realizers. But how can an *instance* of a regularity have a reductive explanation? An instance of a regularity consists of a P-token and then a P*-token, and can have a reductive explanation composed of three elements. The first element is the reductive explanation, of the sort described above, of the existence of the P-token. The second element is a different reductive explanation, also of the sort described above, of the existence of the P*-token. The third element is a *causal* explanation of the *later* non-psychological phenomenon (by which the P*-token was reductively explained) by reference to the *earlier* non-psychological phenomenon (by which the P-token was reductively explained). The three elements taken together constitute an explanation of the psychological regularity: an earlier non-psychological phenomenon both *realizes* (and hence necessitates) the simultaneous P-token and *causes* the later non-psychological phenomenon that realizes (and hence necessitates) the P*-token. So, given the existence of the earlier non-psychological phenomenon and the holding of the non-psychological laws, there *must* exist the P-token followed by the P*-token. And, as reflection will soon show, this explanation of an instance of a psychological regularity clearly meets the conditions for being reductive given in Section 2.1.[12]

3.2 The Character of Psychological Explanation

The second supposedly antireductionist claim was that the explanations that psychologists supply are more than merely heuristic, since they identify objectively existing explanatory factors that the explanations of neurophysiologists (or molecular biologists, or whoever) fail to identify. I'll argue, however, that this claim is quite consistent with psychological reductionism. But I shall do so indirectly, by considering three arguments for fearing that the claim is *in*consistent with psychological reductionism and showing that none of them is sound. The burden of proof will then lie on my opponent to make a positive case for inconsistency.

The first argument holds that if psychological reductionism is true, then psychological phenomena are not real; and that if psychological phenomena are not real, then explanations in psychological vocabulary that cite such phenomena can at best be of heuristic value, since nonexistent phenomena can explain nothing. But this argument fails. Its first premise assumes that if something is reducible, then it doesn't really exist, but the assumption is false. Suppose that you are first acquainted with dry-stone walls from too great a distance to be able to determine their composition; you would surely be wrong to conclude that they don't really exist just because you later discover that they are reducible to the stones that compose them. Any number of other similar examples could easily be produced to show that, at least if our ordinary thought on such matters is correct, reducibility doesn't entail nonexistence. To the contrary, in fact: reducibility entails existence. Since reducibility is a kind of explainability, and since only what exists can be explained, the reducibility of psychological phenomena actually entails that they do exist. So psychological reductionism doesn't require that psychological claims must be dismissed as non-explanatory on the grounds that the phenomena they describe are unreal.

According to a second argument, psychological reductionism entails that all psychological phenomena already have an explanation – a reductive explanation – and hence that there is *no room* for any further explanation such as psychology might supply; so any apparent explanations of psychological phenomena in psychological vocabulary must be treated as heuristic at best. Now this argument clearly rests on the assumption that a single phenomenon can have only one explanation.

But why assume that? Let me consider just one possible reason: if (say) psychological phenomena have multiple explanations, then, implausibly, psychological phenomena will be *causally overdetermined*, since any given psychological phenomenon will have two simultaneous causes – a psychological cause and a non-psychological cause, this latter cause being the cause of the non-psychological phenomenon to which the psychological phenomenon is reducible.

Causal overdetermination, however, is objectionable only if the two simultaneous causes are entirely independent of one another, as they are in the classic examples such as the case where a victim is simultaneously hit by two separate bullets each of which would have sufficed alone to cause death; for only then can one reasonably worry, for example, whether some principle of parsimony (e.g., Ockham's Razor) has been violated. However, given psychological reductionism, the case of a psychological phenomenon that has both a psychological cause and a simultaneous non-psychological cause is importantly different from the two-bullets case precisely because the two causes, unlike the two bullets, are *not* entirely independent of one another. For, given psychological reductionism, the *first* cause of the psychological phenomenon, being itself a psychological phenomenon, must itself be reducible to some simultaneous *non*-psychological phenomenon; and this simultaneous non-psychological phenomenon must *be* (or be *part* of) the *second* cause, i.e., the non-psychological cause of the non-psychological realizer of the psychological phenomenon. Thus, since the first cause *must* (metaphysically, not causally) exist given that the second cause does, it can't be treated as some additional posit in gratuitous violation of Ockham's Razor. So, if psychological reductionism is true, then, precisely *because* it's true, psychological phenomena can have psychological causes *and* simultaneous non-psychological causes, but without overdetermination of the arguably objectionable kind present in the two-bullets case.

There is a variant of the second argument, however, that must also be considered. According to it, psychological reductionism entails that all psychological phenomena already have an explanation – a reductive explanation – and hence that there is no *point* in any further explanation in psychological vocabulary; further explanation isn't impossible, just *needless*. In response to this argument, let me sketch two very abstract ways of thinking about the nature of explanation, both of which are consistent with the idea that multiple explanations of a single phenomenon do have a point. According to the first, to explain some phenomenon is to show that it was necessary (or probable) given some earlier phenomenon; and one can do so by citing some earlier phenomenon and subsuming both it and the explanandum under some law. If so, then a psychological phenomenon can be shown to be necessary, and hence explained, by citing an earlier psychological phenomenon and an appropriate psychological law. But the psychological phenomenon can *also* be shown to be necessary, and hence explained, by citing an earlier *non*-psychological phenomenon which, because of a non-psychological law, causes the non-psychological phenomenon to which the original psychological explanandum is reducible. Neither demonstration has any just claim to be *the* (i.e., the only) explanation of the explanandum. According to the second way of thinking about explanation, to explain some phenomenon is to exhibit it (e.g., by subsuming it under a law) as fitting into some kind of pattern. But if so, then there seems no reason why a single phenomenon should not fit into more than one kind of pattern, in which case we can view a psychological explanation and a reductive explanation of the same psychological phenomenon as exhibiting its fitting into two different, but both entirely real, patterns.

So far in this sub-section I have argued that, even given psychological reductionism, explanations such as psychology supplies cite real phenomena, and are neither impossible nor pointless. But a further question remains, namely, whether such explanations are *indispensable*, not just heuristically but *metaphysically* (see, e.g., Schouten & Looren de Jong, 1999, p. 246). No simple answer is possible. On the one hand, because explanations in psychological vocabulary of psychological phenomena represent those phenomena as fitting into patterns that are fully objective features of the world, whereas reductive explanations of the same phenomena, in a different vocabulary, don't represent them as fitting into *those* patterns but rather into others, explanations in psychological vocabulary really do provide explanatory insight that reductive explanations miss. On the other hand, it must be possible in principle (though not in practice) somehow to use *non*-psychological vocabulary to describe those patterns into which psychological explanations describe psychological phenomena as fitting. For if even this in-principle possibility is denied, physicalism seems to have been abandoned; at the very least, if you endorse physicalism but still think there are contingent items in the world that can't, even in principle, be described using non-psychological vocabulary, then you owe a plausible account of why you still qualify as a physicalist, an account that's precise enough to show that by denying this in-principle possibility you haven't contradicted yourself. However, if non-psychological vocabulary *were* used to describe the patterns into which psychological explanations describe psychological phenomena as fitting, those patterns would be the very *same* patterns into which psychology describes psychological phenomena as fitting; so, the possibility in principle of describing those patterns non-psychologically doesn't detract from psychology's claim to provide explanatory insight.

According to a third argument, psychological reductionism entails that the explanations that psychologists supply aren't genuine because such explanations can't be *causal* explanations. If psychological reductionism is true, then all psychological phenomena are reducible to non-psychological phenomena. But since all non-psychological phenomena have sufficient non-psychological causes, it seems to follow that among psychological phenomena no genuine causal relations hold, even though they appear to. Jaegwon Kim has long defended this kind of argument, though in his formulation a supervenience thesis plays the role played here by psychological reductionism (see, for his most recent version, Kim 2005, ch. 2).[13] But in fact the conclusion of this argument doesn't follow from its premises; the so-called "causal exclusion principle" on which it relies seems to be just false. Consider our judgment of an analogous case. Suppose that as I walk beside a large dry-stone wall it falls, bruising me extensively; clearly my extensive bruising was caused by the falling wall. Now, because the wall is entirely made up of stones, the falling wall is reducible to many falling (and doubtless interacting) stones; similarly, the extensive bruising is reducible to many sub-areas of bruising. Now, we could in principle give a sufficient explanation of every single sub-area of my extensive bruising by reference to the career of some or other particular falling stone. But our ability to do so, and the reducibility which makes it possible, cast no doubt at all on our original judgment that my extensive bruising was caused by the falling wall. At least if our ordinary thinking about the matter can be trusted, the causal explanation of my extensive bruising that cites the falling wall can peacefully coexist with the "reductive" causal explanation that cites the many falling stones that individually cause the many sub-areas of bruising that constitute my extensive bruising. Likewise, I assume, for the case of causal explanations framed in psychological and in non-psychological terms: causal explanations of both kinds can coexist even if psychological reductionism is true.[14]

3.3 The Methodological Autonomy of Psychology

The third supposedly antireductionist claim was that it is possible for psychologists to discover psychological entities, psychological processes, and the psychological generalizations that govern them without either investigating or knowing about the underlying physical (e.g., neurophysiological or biochemical) phenomena. However, psychological reductionism doesn't rule out this possibility. To see why, it will help to explore more deeply what psychological reductionism is, and what it is not.

What it is is a thesis, with a truth-value, concerning the relationship between the property-instances, events, objects, and so on (correctly) described by psychology and the property-instances, events, objects, and so on (correctly) described by sciences other than psychology. If true, it contributes to addressing a larger question in which everyone curious about what the world is like – and hence, one would have thought, every scientist – should be interested. on the basis of the evidence currently available to us, what is the relationship between the various segments of reality investigated by the many sciences? Suppose someone asks us what the world – at least the contingent world – is like, and that, as good naturalists, we respond by reading out the contents of all of our science textbooks, adding that this answer, though certain to be wrong in many respects, is nonetheless the best we presently have. There is still a question as yet unanswered, concerning how the various accounts of the world given in the various textbooks fit together to form a unified account of the contingent world as a whole. It is to this question that psychological reductionism offers part of an answer. Accordingly, both the question and the answer can be said to belong to metaphysics if metaphysics is the part of philosophy that aims to understand what the world is like, but at a higher level of abstraction than that afforded by particular branches of science – but to metaphysics of a kind that philosophers of science should welcome.

What psychological reductionism is not is a thesis about the relationship between certain *theories* (e.g., psychological ones and neuroscientific ones), either *at* a time or *over* time, although an account of inter-theoretic reduction for true theories could be constructed from it.[15] Nor is psychological reductionism a thesis about the actual relationship between psychology as an *activity* or *practice* and other sciences considered as *activities* or *practices*. The nature of this relationship is certainly a fascinating question for sociologists and even philosophers of science, but it is not the question to which psychological reductionism is addressed. Finally, psychological reductionism, being a *descriptive* thesis, is not a methodological directive, which would be *prescriptive*.

Once the character of psychological reductionism has been properly understood, I think it becomes clear that it doesn't entail *anything* about how discoveries in psychology can or cannot be made. A fortiori, it doesn't entail that psychologists can't make psychological discoveries unless they investigate or know about the non-psychological phenomena to which psychological phenomena are reducible. In further support of this contention, consider the analogous case of water, which is presumably reducible in the present sense to chemical entities. There is much that could have been discovered about the macro-features of water (e.g., its viscosity, specific heat, boiling and freezing points, necessity for life, role in chemical reactions) by someone who neither investigated nor knew about the atoms of hydrogen and oxygen to which it is reducible. The position is no different, I suggest, for reducible psychological phenomena.

3.4 Top-down Correction

The fourth supposedly antireductionist claim was that it is possible for discoveries in psychology to *correct* claims made by those who study the underlying physical phenomena (see, e.g., Schouten & Looren de Jong, 1999, pp. 256–258). This possibility, too, is quite consistent with psychological reductionism, as you would expect given that psychological reductionism doesn't even *aim* to comment on the relationship between the *practice* of psychology and that of other sciences. The puzzling thing is why anyone would have expected psychological reductionism to rule out this possibility. The answer, I suggest, is that psychological reductionism indeed entails that certain non-psychological phenomena are *metaphysically* prior to psychological phenomena, and from this someone might infer that the relevant non-psychological sciences are therefore *epistemologically* prior to psychology, with correction flowing only upwards, from non-psychological claims to psychological ones. But such an inference would be unsound, as I'll now explain.

Imagine embarking on a particular psychological inquiry while assigning a significant probability to psychological reductionism – perhaps on the ground that other psychological phenomena, and other common-sense but non-psychological phenomena, have as a matter of fact turned out to be reducible in the pertinent sense. You are therefore confident that the psychological phenomena you are aiming to discover do *in fact* mesh in a certain way with (i.e., are in fact reducible to) whatever the relevant simultaneous non-psychological phenomena turn out to be. For this reason, it counts against a proposed psychological hypothesis if the phenomena it postulates *can't* mesh with (what our best theories claim to be) the relevant non-psychological phenomena; it doesn't, of course, *conclusively falsify* the psychological hypothesis, but it does count against it. But for the very same reason, it also counts against a relevant *non*-psychological hypothesis if the non-psychological phenomena that *it* postulates can't mesh with (what our best psychological theories claim to be) the relevant *psychological* phenomena. Just as the assumption of psychological reductionism makes it reasonable for correction to flow *upwards*, from non-psychological to psychological claims, so also the same assumption makes it reasonable for correction to flow *downwards*, from psychological to non-psychological claims.

In fact, it is hard to see how anything *other* than some sort of metaphysical reductionism could underwrite the reasonableness of the two-way flow of correction between psychology and the relevant non-psychological sciences. For suppose that psychology and, say, the neurosciences investigated domains that were entirely independent of one another except for the holding of certain fundamental cross-scientific laws; this is the radically non-reductionist picture of the relations between psychology and the neurosciences that psychophysical dualists must favor.[16] Then, presumably, psychology and the neurosciences would – and should – develop quite independently of one another, except for when it was necessary to discover such fundamental laws as might hold between their respective domains. Precisely because psychological phenomena were *not* reducible to non-psychological phenomena, there would be no rationale for testing psychological hypotheses by examining neuroscientific phenomena, or vice versa.

The methodological claim that psychology and the neurosciences should co-evolve would therefore appear to *presuppose* some version of metaphysical reductionism.

Notes

1 There *may* be a form of dualism that holds that psychological properties, and perhaps even minds themselves, are *metaphysically* dependent on the physical properties of brains and their physical environments (see Melnyk, 2003, pp. 58–62).

2 The picture is *broadly* functionalist because it isn't committed to any *particular* hypotheses about the functional character of psychological phenomena.

3 *Contra* Huib Looren de Jong, who holds that functionalism can't allow the possibility of biological constraints on psychological theorizing (Looren de Jong, 2002, p. 457).

4 I shan't here argue that this form of psychological reductionism is true, but evidence for it is, in effect, presented in (Melnyk, 2003, ch. 6).

5 Consensus theories of current psychology are those that are currently the objects of consensus among scientific psychologists. A good way of determining which theories these are is to examine the contents of widely used textbooks in psychology.

6 Another possible reason is this: in specifying which phenomena should count as the neurophysiological ones, we would need to appeal to what is asserted to exist by neurophysiological theories; but in that case we would run into familiar difficulties concerning whether the intended theories belong to current or to future (or even to ideal) neurophysiology.

7 Thus, the possession by psychological phenomena of biological functions is not in itself a reason to doubt psychological reductionism, *pace* Schouten & Looren de Jong, 1999, pp. 240–246.

8 Note that premises (1) and (2) needn't, and typically won't, describe the *totality* of non-psychological phenomena simultaneous with the tokening of P. The descriptions they give will be *selective*, confining themselves to relevant non-psychological phenomena, i.e., those relevant to the realization of a token of P. Therefore, the non-psychological phenomena cited in a psychological reduction (if it takes the form sketched in the text) can't be charged with indiscriminately including irrelevant non-psychological detail. And no obstacle to reduction is created by "multiple supervenience," i.e., "the capacity of one and the same physical substrate to support a number of different higher-level

properties" (Schouten & Looren de Jong, 1999, p. 241; see also Ross & Spurrett 2004, Sect. 3.2), since reductions of (tokens of) different higher-level properties will cite different *aspects* of the same physical substrate.

9 Here, I am just assuming that this derivation represents a genuine explanation. For defense, see (Melnyk, 2003, pp. 98–102).

10 I take this point from Jaegwon Kim (Kim, 2005, pp. 131–9).

11 This claim is highly contentious. For defense, see (Melnyk, 2003, pp. 104–109).

12 The extent of a psychological regularity can perfectly well be restricted spatio-temporally, as Looren de Long plausibly suggests is true of biological regularities (Looren de Jong, 2002, p. 453). However, restricted regularities can be reductively explained along the lines I've sketched – by explaining each of their instances – just as well as unrestricted regularities can.

13 By the way, Kim wouldn't regard psychological reductionism as strong enough for true reductionism, since it doesn't entail the holding of token or type identities between psychological and non-psychological phenomena.

14 For an account of causation that tries to show *how* this coexistence is possible, see (Melnyk, 2003, ch. 4).

15 Perhaps this: theory T_1 is inter-theoretically reducible to theory T_2 if the ontology of T_1 is metaphysically reducible to the ontology of T_2.

16 At least those psychophysical dualists who don't hold that a science of the mind is impossible in principle.

References

Dretske, F. (1988). *Explaining Behavior: Reasons in a World of Causes.* Cambridge, MA: MIT Press.

Fodor, J. (1974). Special sciences, or the disunity of science as a working hypothesis. *Synthese*, *28*, 97–115. Reprinted in N. Block (Ed.), *Readings in Philosophy of Psychology* (Vol. 1, pp. 120–33). Cambridge, MA: Harvard University Press, 1980.

Fodor, J. (1990). *A Theory of Content and Other Essays.* Cambridge, MA: MIT Press.

Kim, J. (2005). *Physicalism, or Something Near Enough.* Princeton: Princeton University Press.

Looren de Jong, H. (2002). Levels of explanation in biological psychology. *Philosophical Psychology*, *15*, 441–462.

Lycan, W. (1987). *Consciousness.* Cambridge, MA: MIT Press.

Melnyk, A. (1995). Two cheers for reductionism; or, the dim prospects for non-reductive materialism. *Philosophy of Science*, *62*, 370–388.

Melnyk, A. (2003). *A Physicalist Manifesto: Thoroughly Modern Materialism*. Cambridge: Cambridge University Press.

Millikan, R. G. (1984). *Language, Thought, and Other Biological Categories: New Foundations for Realism*. Cambridge, MA: MIT Press.

Putnam, H. (1975). *Mind, Language, and Reality: Philosophical Papers* (Vol. 2). Cambridge: Cambridge University Press.

Ross, D., & Spurrett, D. (2004). What to say to a skeptical metaphysician: A defense manual for cognitive and behavioral scientists. *Behavioral and Brain Sciences*, *27*, 603–627.

Schouten, M., & Looren de Jong, H. (1999). Reduction, elimination, and levels: The case of the LTP-learning link. *Philosophical Psychology*, *12*, 237–262.

3

SOME METAPHYSICAL ANXIETIES OF REDUCTIONISM

Thomas W. Polger

1. Introduction

By now it is a cliché to observe that so-called reductionism is not one mammoth doctrine. There are, as it were, many reductionisms. Needless to say, there are at least as many antireductionisms. Despite the fact that neither reductionisms nor their counterparts are single and unified doctrines, there do seem to be some family resemblances. One, it seems to me, is that both reductionisms and antireductionisms are acute responses to certain metaphysical worries. Some of these worries are metaphysical in nature, and others are worries about the nature of metaphysics. My contention is that these worries are by and large misguided, and thus that the anxious reactions of both reductionists and antireductionists are unwarranted.

For the present purposes I will distinguish between reductionist and antireductionist *theses*, on the one hand, and reductionist and antireductionist *approaches*, on the other. This is a perhaps clumsy distinction, and I don't know that it carves reductionism at its joints. But I think it can be made to do some work. By *theses* I have in mind particular views about the nature of reduction and reductive relations, which can be worked out in various ways, some of which will be discussed below. By *approaches* I have in mind the motivations and background assumptions that go into formulating or adopting particular theses. Individual reductionists and antireductionists usually hold what we might then call a *theory*, a combination of an approach and a thesis. A great deal has been written about the merits of or truth of any number of reductionist or antireductionist theses. Presumably there are some assumptions about approaches and motivations lurking in the background. But very little has been said about the merits of the various approaches themselves. Put another way: There is a lively debate over whether particular reductionist (or antireductionist) theses can solve certain problems or avoid certain objections. But not enough attention has been given to whether these are problems worth solving or objections worth overcoming. If the approaches are based on faulty assumptions, as I will argue, then they may misrepresent the available alternatives.

In this chapter I characterize four general approaches: Metaphysical Reductionism, Metaphysical Antireductionism, Antimetaphysical Reductionism, and Antimetaphysical Antireductionism. In each case I will offer some paradigm examples; my aim is to give representative samples rather than a comprehensive catalog. (So if there are some approaches that do not fit my framework I will not be bothered. On the other hand, if no views fit my model, that would be a problem.) In the process of distinguishing these approaches and presenting examples, I will make the case that each originates in a reaction to a metaphysical (or antimetaphysical) concern. Then I will argue that all four approaches accept an assumption that I call the *autonomy thesis*. If the autonomy thesis can be rejected, then we need not choose among the four approaches. Reductionism and antireductionism are not the only ways of resolving our metaphysical concerns.

I'll begin by motivating this new taxonomy of reductionist and antireductionist approaches.

2. Ontological and Theoretical Reductionisms

Before introducing my new taxonomy of reductionisms, it is useful to remind ourselves of what I aim to replace. The usual way to begin classifying reductionisms is by distinguishing those with ontological theses from those with explanatory or theoretical theses.[1] Ontological reductionism is a thesis about the furniture of the world. It says that we only need to accept into our ontology some limited set of "basic" entities, properties, events, or processes. Usually this basic set is supposed to be provided by physics (strictly speaking) or the physical sciences (broadly construed.) The ontological reductionist is usually contrasted with the explanatory or theoretical reductionist. Theoretical reductionism is a thesis not about the relations among objects, but about the relations among explanations, theories, or sciences. It says that the explanatory or predictive work that is done by some "higher-level" theory can be done by a different "lower-level" theory. Thus the higher level theory is said to be reduced to, translated to, or replaced by the lower-level theory.

This traditional contrast between ontological and theoretical reductionisms has some virtues. Chief among those virtues is that it allows us to see how the ontological and theoretical theses can come apart. For while adherence to theory reduction is a reason to adopt ontological reduction (and vice versa), it is also possible to hold the ontological thesis while rejecting the theoretical thesis.[2] Indeed this combination has come to be the received view in philosophy of mind and philosophy of science, where it is often called nonreductive physicalism. (I'll prefer the expression "antireductionist" because it expresses the positive thesis that some phenomenon is not reducible better than the faux agnosticism of "nonreductive.")

The downside of drawing the distinction in this way is that it has seemed to license a certain willful negligence of problems that originate on the "other" side of the divide. For example, those concerned with theory reduction may maintain that their theses are entirely insulated from ("autonomous from") ontological worries about the causal powers of higher level entities. Similarly, ontological reductionists may have little interest in shifting views as to what counts as an explanation in an "autonomous" science that ignores their ontological burdens. It seems that this framework has become an obstacle to understanding reduction, so it is time for a fresh perspective.

3. Metaphysical Anxieties and Reductionisms

There are two kinds of broadly metaphysical concerns that, it seems to me, drive most reductionist and antireductionist approaches. The first are classic problems within metaphysics, concerning causation, material composition, emergence, natural properties, and so forth. The second are concerns about the utility of metaphysics as an endeavor, about the meaningfulness of metaphysical claims, or the justification for ontological commitments. These two sorts of concerns can motivate both reductionists and antireductionists.

Before characterizing the four approaches that are my subject herein, some things must be said about "reductionism" in general. First, we should understand some common ground in the very idea of reduction. Reductionists and antireductionists agree that nature or science consists of some number of levels of objects or theories. One point of contention that crosscuts the reductionist/antireductionist divide concerns what it is to be or to be located in a level. (As Richardson has noted: Even eliminative reductionists believe in levels – they simply believe that there is only one of them!) Reductionists and antireductionists agree that levels of objects or theories are organized hierarchically by something like a containment relation: if there are higher-level objects or theories, then those are made or constructed out of lower-level objects or theories. Lower-level objects are parts or realizers of higher-level objects; lower-level theories are more basic than or more logically primitive than higher-level theories. A second point of contention that is orthogonal to the reductionist/antireductionist divide concerns whether there is any absolute "bottom" level of ontology or explanation, or for a particular reductive claim which (relatively) lower level of ontology or explanation should be taken to be the reductive base. Typically the lower-level explanations or objects are said to belong to (in decreasing order of austerity) physics, physics and chemistry, inorganic sciences, or non-intentional sciences, and their respective ontologies. Finally, reductionists and antireductionists agree that there is some relation R that, if it held, would count as the "reduction" of a disputed ontology or theory to a more basic or absolutely basic ontology or theory. Candidates for the reduction relation R are myriad. Some of the most common accounts maintain that H is reducible to L if and only if: (a) H is (or facts about H are) logically derivable from L (or facts about L); (b) H is conceptually entailed by L; (c) H is nomologically necessitated by L; (d) H is nomologically necessitated by L and some bridge laws; (e) H can be exactly modeled in L; (f) H can be approximately modeled in L; (g) H can be replaced by L; (h) H is realized by L; (i) H is identical to L; (j) H can be transparently explained by L. The options on this list are neither exclusive nor exhaustive, but they ought to be familiar.

The bottom line is that any particular reductionist or antireductionist thesis will be cast in terms of some notion of levels, some notion of a base level, and some reductive relation. This much is common ground for reductionists and antireductionists, and much ink has been spilt over whether there is a correct or ideal variation, and whether (in any particular debate) the disputants are talking past one another by assuming different variations.

The second thing to be said about reduction in general is that there is variation among reductionist and antireductionist theses as to what is the reductive target, about the ontology or theory that is said to be reducible or irreducible. Among the most familiar targets are intentional states or explanations, phenomenal properties or facts, organic entities and biological explanations, genes and genetic theory, and everything at all. If we are interested in reduction generally, then the particular reductive target will generally be less crucial than the other variables.[3]

The final thing that must be said is that my discussion herein will abstract away from such details. This is because I focus on what I am calling reductionist and antireductionist approaches rather than detailed theses. To say this is not to say that the details of the theses do not matter to the approaches, for indeed they do matter. Rather I am suggesting that there are interesting similarities and differences among reductionist and antireductionist approaches that can be profitably explored while bracketing these in-house disputes. I will usually take it for granted that the reader will see how the particular theses may modulate the approaches, but in a few cases I will note these connections directly.

So there are, on the one hand, numerous particular reductionist and antireductionist theses that can be formulated by making choices about levels, reductive bases, and reductive relations. My present concerns, on the other hand, are four basic approaches to the question of reductionism that (I maintain) can be fruitfully explored without first selecting a particular reductive or antireductive thesis. It is to these four approaches that we now turn.

3.1 Metaphysical Reductionism

I am distinguishing between approaches to reduction (and antireduction) that are motivated by concerns that are metaphysical in nature and those that are motivated by concerns about the nature of metaphysics. Metaphysical Reductionism is the approach that advocates for reductive theses for reasons that are broadly metaphysical in nature. Of course I offer no account of what makes a concern broadly metaphysical, but I don't think we need to resolve that demarcation problem in order to recognize some examples.

Jaegwon Kim's causal exclusion argument is a clear case of a metaphysical approach to reduction (1993, 1998, 2005). Kim argues that higher-level properties are either causally redundant or else causally inert. Discounting general causal overdetermination, they are causally epiphenomenal. But, by an ontological principle that Kim calls Alexander's Dictum, to exist is to have causal powers. Since higher-level properties have no causal powers of their own, the only way to save higher-level properties (and their bearers) is to reduce them to respectable lower-level properties that are the genuine locus of causal powers. Most recently Kim has argued that this reduction must proceed by construing the higher-level properties as functional properties that are realized by the causally potent lower-level properties (1998, 2005). That is, Kim endorses a particular way of formulating the reductive thesis that is motivated by the causal exclusion argument. But for present purposes the crucial bit is that Kim's general approach is driven by metaphysical or ontological concerns, namely the dedication to a certain ontology and a general ontological principle.

A somewhat different strand of Metaphysical Reductionism is represented by the analytic reductionism revived by David Lewis (1994), David Chalmers (1996), and Frank Jackson (1998). This approach has garnered much more attention in philosophy of mind and metaphysics than in philosophy of science. According to Jackson, "serious metaphysics" requires accounting for everything that there is in terms of some relatively small privileged class of objects. This accounting problem is what Jackson calls the "location" problem, for all objects must be located vis-à-vis the privileged ontological class. More precisely, the truth makers for the H-facts must be located among the truth makers for the L-facts. Jackson holds that this requires reducing the H-facts to the L-facts, and that in turn requires that the H-facts be a priori "entailed" by the L-facts. The particular example of a serious metaphysical view with which Jackson concerns himself is physicalism, the view that the truthmakers for all facts can be located among the truthmakers for the strictly physical facts. Reversing his well-known previous dualism (1982), Jackson now believes that physicalism is true.

Jackson's physicalist reduction counts as a Metaphysically Reductionist approach for two reasons. First, Jackson's reason for thinking that physicalist reductionism must be true is that he now finds epiphenomenalism unacceptable, and he believes that any nonphysicalist or nonreductive view will make the irreducible properties epiphenomenal. In short, he accepts Kim's causal exclusion argument. Second, Jackson's "serious metaphysics" is itself a reductive enterprise that is based on explicitly metaphysical principles. As Jackson explains it, any serious metaphysics will locate all facts and entities (that it recognizes) among a privileged class of facts and entities. This is a metaphysical injunction that is stronger than Ockham's Razor. Ockham's Razor says that we should not multiply entities beyond necessity. Jackson's serious metaphysics assumes that we should not multiply entities, period. This is more than mere dedication to desert landscapes. It is unclear on what grounds a serious metaphysician could justify her selection of the privileged entities and facts. But it is clear that they cannot be ordinary Ockhamist grounds for those do not come into play until after the basic entities and facts have been established. And it cannot be, as for Quine, that we look to science; for that will only give us the sets of facts that need locating. (The sciences themselves give us only a "Big List" ontology, as Jackson might say.) The important point, for our purposes, is not that Jackson could never give reasons for picking, e.g., physical facts as the base facts. Rather, our interest is in noticing that Jackson's choice will be a metaphysical one, and so provides a second reason for thinking of his reduction as a kind of Metaphysical Reductionism.

3.2 Antimetaphysical Antireductionism

Antimetaphysical Antireductionism is the approach to reduction paradigmatically represented by Jerry Fodor's "Special Sciences" article (1974) and Hilary Putnam's example of the square peg and round hole (1975). These two papers, along with Kitcher (1984), are responsible for making "nonreductive physicalism" the received view in philosophy of mind and philosophy of science.

The approaches of Fodor, Putnam, and Kitcher are antimetaphysical in the sense that they are suspicious of privileging metaphysical concerns. Instead they put explanatory concerns in the driver's seat and "let the ontological chips fall where they may" (Bickle, 2003, p. 32).[4] For this reason, it is tempting to think that Fodor, Putnam, and kindred spirits are concerned with what the usual taxonomy calls explanatory or theoretical reduction, as contrasted with the ontological reduction that occupies Kim, Jackson, and their ilk. It would then be tempting to defuse their apparent disagreement by observing that one group is concerned with ontology while the other is concerned with explanation, and suppose that they are simply talking past one another. Indeed, I believe that this is more or less the standard view. But this view is incorrect, and it goes wrong because it undercounts the degrees of freedom in the debate over reductionism. The two groups may indeed be talking past one another, but it is not because they are simply concerned with distinct questions. Rather, it is because their approaches differ doubly: on the truth of reduction, and on the approach to or motivation for reduction.

Fodor famously argues that, *pace* Oppenheim and Putnam (1958), the multiple levels of sciences should be thought of as disunified. It may be that occasional unifying reductions can be had. But the "working hypothesis" is that they need not be found in order to legitimate special science (i.e., higher-level) explanations. On Fodor's view, reductionism is the very unlikely empirical hypothesis that every kind corresponds to a physical kind:

> The reason it is unlikely that every kind corresponds to a physical kind is just that (a) interesting generalizations (e.g., counterfactual supporting generalizations) can often be made about events whose physical descriptions have nothing in common; (b) it is often the case that whether the physical descriptions of the events subsumed by such generalizations have anything in common is, in an obvious sense, entirely irrelevant to the truth of the generalizations, or to their interestingness, or to their degree of confirmation, or, indeed, to any of their epistemologically important properties; and (c) the special sciences are very much in the business of formulating generalizations of this kind. (1974, p. 124)

So the doctrine of reductionism is false not because reductions are impossible, but because they are unnecessary. The popularity of reductionism, according to Fodor, owes to its confusion with the doctrine that physics is the most basic and most general science, and with the doctrine that every token of a special science kind is identical to a token of a physical kind, i.e., token physicalism. Fodor's antireductionism is antimetaphysical because it is motivated by the availability of higher-level lawlike generalizations irrespective of their ontological relation to lower-level generalizations. In the same vein, Kitcher argues that molecular details are simply irrelevant to explanations of in terms of chromosomes (1984, p. 348).

Putnam (1975) advances a similar view using the example of a square peg and round hole. Describing a board with holes cut in it, one round and one square, and a square peg, he poses the question: "We have the following very simple fact to explain: *the peg passes through the square hole, and it does not pass through the round hole*" (1975, p. 295). The correct or best explanation, he insists, is an explanation of the higher-level entities and phenomena:

> the board is rigid, the peg is rigid, and as a matter of geometrical fact, the round hole is smaller than the peg, the square hole is bigger than the cross-section of the peg. The peg passes through the hole that is large enough to take its cross-section, and does not pass through the hole that is too small to take its cross-section. (1975, p. 296)

This higher-level explanation is superior to any lower-level explanation because, among other merits, the higher-level explanation can be applied to other geometrically similar pegs and holes that are made of different materials and so is more general than the lower-level explanation. It is the higher-level features that are relevant to the event and its explanation.[5]

Like Fodor and Kitcher, Putnam does not doubt that token pegs, genes, or psychological states are physical entities. But that fact is simply not relevant to the content of or the legitimacy of higher-level explanations. In this sense, higher-level explanations and sciences are "autonomous" from physics and other lower-level explanations and sciences. We may say about pegs and genes, as Putnam says about mental states, that it does not matter whether they are made of "copper, cheese, or soul" (1975, p. 292). Ontology is secondary to the explanatory goals, and those do not depend on reduction, according to the Antimetaphysical Antireductionists.

3.3 Metaphysical Antireductionism

Thus far I have not been concerned with evaluating reductionist and antireductionist approaches, only with stating them. But to understand the Metaphysical Antireductionist approach it is useful to recognize two difficulties that face its antimetaphysical counterpart, Antimetaphysical Antireductionism. First, the autonomy of special sciences from lower-level sciences can come into tension with the truth of special science explanations. Putnam himself is candid about this problem. Speaking of the autonomy of Newtonian mechanics from contemporary gravitational physics, Putnam writes, "I want to draw the philosophical conclusion that Newton's laws *have a kind of reality in our world* even though they are not *true*" (1975, p. 301). If one is at all inclined toward even mild forms of scientific realism or toward conceptions of explanation as ontologically committed, then this tension will be a genuine problem.[6] Second, the autonomy of the special sciences can come into tension with some explanatory claims such as causal claims. If both the autonomous upper-level sciences and the lower-level base sciences make causal claims, then questions will arise about the relations between the various causal claims. In short, the kinds of causal exclusion worries that motivate Kim's Metaphysical Reductionism will have to be dodged or answered. The upshot is that any particular Antimetaphysical Antireductionist thesis will have to offer a way out of these tensions. And this has proven difficult.[7]

As I already noted, Putnam explicitly raises the problem about truth. It seems to me that Fodor, Putnam, and Kitcher all recognized the potential instabilities with the Antimetaphysical Antireductionist position, and each of them is tempted to avoid the above tensions by taking a stronger stand. They each gave in to this temptation in the very same articles discussed above, in which their official doctrines were in every case Antimetaphysical Antireductionist approaches. The stronger stand is to assert that not only is reductionism not necessary for the legitimacy of the special science explanations, but that reduction is impossible because of the actual structure of the world. This stronger position is a version of Metaphysical Antireductionism because it maintains that there are entities, events, properties, or facts in the world that cannot be recognized by reductionists. Fodor, for example, writes,

> I am suggesting, roughly, that there are special sciences not because of the nature of our epistemic relation to the world, but *because of the way the world is put together*: not all the kinds (not all the classes of things and events about which there are important, counterfactual supporting

> generalizations to make) are, or correspond to, physical kinds . . . Physics develops the taxonomy of its subject matter which best suits its purposes: exceptionless laws which are basic in the several senses discussed above. But this is not the only taxonomy which may be *required* if the purposes of science in general are to be served: e.g., if we are to state such true, counterfactual supporting generalizations *as there are to state*. (1974, p. 131, emphasis added)

The idea suggested by Fodor in this passage, and that is characteristic of the Metaphysical Antireductionist view, is that there are patterns in the world that simply cannot be noticed or stated in terms of lower-level entities or generalizations. Such entities or generalizations are irreducible (and thus autonomous) because higher-level terms and explanations are necessary ("required") in order to express them. On this view, the lower-level stuff and the possibility of reduction are *not* irrelevant to the legitimacy of higher-level entities and explanations. Quite the opposite, it is the failure of reduction due to the incompleteness of lower-level ontology and explanation that justify the higher-level explanations and their ontologies.

Putnam, similarly, flirts with Metaphysical Antireductionism when he suggests that, in the case of the square peg and round hole, the microphysical explanation is not an explanation at all and the macrophysical explanation is "correct" (1975, p. 296). Recognizing the strangeness of this claim, he explains:

> People have said that I am wrong to say that the microstructural deduction is not an explanation. I think that in terms of the *purposes for which we use the notion of explanation*, it is not an explanation. If you want to, let us say that the deduction *is* an explanation, it is just a terrible explanation, and why look for terrible explanations when good ones are available? (1975, p. 296, original emphasis)

The idea that Putnam is flirting with is that the lower-level explanation is not an explanation at all of the phenomenon of interest, that the phenomenon of interest *must* be explained at the higher level. If it must be explained at the higher level, then it cannot be reduced to any lower-level explanation. The suggestion is that the phenomenon will simply be missed by or invisible to the lower-level viewpoint. As such, the lower-level explanation is simply inadequate and reduction is ruled out.

Kitcher, too, leans on the stronger thesis. Like Putnam, he sometimes suggests that the lower-level explanations are simply explanatorily inadequate: "appeal to molecular biology would not deepen our understanding of the transmission law [of chromosomes] . . . In charting the details of the molecular rearrangements the derivation would only blur the outline of a simple cytological story, adding a welter of irrelevant detail" (1984, p. 347). But Kitcher also implies that there are facts in the world that a lower-level account will simply miss: "The molecular derivation forfeits something important" because the cytological processes "cannot be identified as a kind from the molecular point of view" (1984, p. 349). Thus, "[t]he molecular account objectively fails to explain because it cannot bring out that feature of the situation which is highlighted in the cytological story" (1984, p. 350).

So Fodor, Putnam, and Kitcher are tempted by the Metaphysical Antireductionist approach in some of their officially Antimetaphysical Antireductionist works. But others assert this stronger thesis outright. Louise Antony and Joe Levine write, "a property is real (or autonomous) just in case it is *essentially* invoked in the characterization of a regularity" (Antony & Levine, 1997, p. 91). Antony and Levine argue that mental states and properties are both reducible and autonomous. Autonomy is usually understood by contrast with reduction, as per Fodor: "Simply to have a convenient way of talking, I will say that a law or theory that figures in bona fide empirical explanations, but that is not reducible to a law or theory of physics, is ipso facto *autonomous*" (1998, p. 149). So clearly Antony and Levine must have different reductionist theses in mind when they endorse one and deny another to arrive at reduction with autonomy. For the present purpose, however, I want to focus on the autonomy side of the claim, and on the approach to antireductionism that leads them to it. The answer, it seems, is that Antony and Levine are led to the autonomy thesis via metaphysical concerns, and so that aspect of their view fits the Metaphysical Antireductionist picture. The reason is that they assert the irreducibility of mental states in order to preserve the reality of those states and properties. For, by the principle quoted above, if a mental state or property is not essentially – irreducibly – invoked in a generalization at some level, then it is not a real state or property at all. That is, they see reduction as a route to elimination, so the motivation for asserting antireductionism is to preserve the higher-level ontology.[8]

A quite different flavor of Metaphysical Antireductionism can be found in Ned Block's and Robert Stalnaker's response to the analytical reductionism that Chalmers (1996) and Jackson (1998) suppose must be endorsed by any physicalist. On one reading, Block and Stalnaker (1999) can be seen as arguing that it is not true that all facts are reducible (in the a priori analytic way required by Chalmers and Jackson) to strictly physical facts, on the grounds that there are some a posteriori but metaphysically necessary facts that are not so reducible.[9] Among these, they think, are that water is identical to H_2O and that water boils at 100°C. This approach is metaphysical because it is motivated by ontological commitment to objects and facts other than those that belong to strict physical theory.[10]

On the Metaphysical Antireductionist approach, there are generalizations about higher-level entities and properties that literally cannot be seen, characterized, or captured at the lower level. These higher-level generalizations are, therefore, indispensable. And their existence means that reduction must be false because the world does not cooperate with the reductionist paradigm. This is a view that Schaffner (1993) calls "in-principle emergence" and Chalmers (1996) calls "dualism."

3.4 Antimetaphysical Reductionism

According to Antimetaphysical Reductionists, reductionism is true but it is motivated entirely by scientific or metascientific concerns. From this approach, it will look as though any antireductionist is appealing to suspicious metaphysical principles to justify questionable ontological commitments that can be shown by empirical means to be either false or untestable.

Paul and Patricia Churchland are perhaps the most often cited representatives of Antimetaphysical Reductionism. They have argued that intentional states (P. M. Churchland, 1981, 1982) and conscious states (P. S. Churchland, 1983) should be eliminated because nothing that science has discovered has the characteristics that intentional or conscious states are supposed to have, explanations made at the psychological level are false, and better explanations can be given at lower levels, viz., the levels of the neurosciences.

John Bickle (1998) took a similar view, arguing that on a continuum of smooth (little revision of the reduced theory required) to rough (much revision of the reduced theory required) reductions, psychological states get relatively rough reductions.[11] But Bickle now regards his earlier (1998) view as too metaphysical:

> Even though scientists talk a "realistic"-sounding language, we should not interpret this talk as addressing questions "external to" the practices and concerns of a given scientific endeavor. The job of new wave metascience is simply to illuminate concepts like reduction as these imbue scientific practice. To what end? Not to achieve some better way of addressing reformulated "external" questions about the existence and nature of "theory-independent ontology." Rather, because a reasonable explanatory goal is to understand practices "internal" to important current scientific endeavors and the scope of their potential application and development. The tasks of this book are part of a metascience of contemporary psychology and neurobiology, not a part of some "ontology of mind." (2003, p. 32)

So Bickle's new wave metascience is suspicious of any metaphysical or ontological claims, drawing instead on Carnap's (1950/1956) distinction between internal and external questions. Bickle joins Carnap as dismissing ontological or metaphysical questions as external to scientific practice and thereby literally meaningless. His reductionism, then, is itself a claim internal to science; and in particular it is supposed to be internal to the practices of molecular neuroscientists. (Better: It is supposed to be a description of the internal practices of molecular neuroscientists. It need not be part of their practices to spend time describing their practices, though of course it could be.)

Whether Bickle is correct is not our present interest. We are concerned with his antimetaphysical approach to reduction. It is antimetaphysical because it insists that reduction is an internal thesis, and it is reductionist then in an entirely descriptive mode. Bickle, in essence, tells us, "This is what [a certain group of] neuroscientists do."

3.5 Hard to Classify

These four options seem to me to reasonable organize the field of contenders (Figure 3.1). I said from the start that my taxonomy was meant to be useful but not canonical, and that some views may not be cleanly pigeonholed by this way of thinking. Before going on, I'll take a moment to point out one such view and indicate why it

	Reductionist	Antireductionist
Metaphysical	Kim, Jackson, Chalmers, Lewis	Fodor, Putnam, Kitcher, Antony & Levine
Antimetaphysical	Carnap, P. M. Churchland, P. S. Churchland, Bickle	Fodor, Putnam, Kitcher, Rudder Baker

Figure 3.1 Four approaches to reductionism and antireductionism, and some paradigm advocates of the approaches

is not obvious where to situate it. By giving specific reasons for the lack of fit, I hope to assuage worries that the problem is with my own taxonomic principles.

Consider Daniel Dennett's "stances" approach to intentional explanation (1978, 1987). On this view, roughly, for a system to have intentional states is for its behavior to be usefully explained and predicted by treating the system as approximating optimally rational behavior. The utility of the intentional explanation is quite independent of whether any other explanations (particularly, for Dennett, physical or design stance explanations) are available. Thus Dennett's view is clearly antireductionist. But is it an example of Antimetaphysical Antireductionism or of Metaphysical Antireductionism? On first pass, the stances approach seems to be antimetaphysical, giving priority to explanatory and predictive concerns. But matters are not so simple. For on Dennett's view, the question of whether a system can be treated as an intentional system and whether it "really is" an intentional system are the same. Having decided whether the intentional stance is useful, no further question remains regarding whether a system is, metaphysically speaking, an intentional system.

Simply put, Dennett's instrumentalism or quasi-realism rejects the kind of distinction between metaphysical and antimetaphysical approaches that I am employing. A similar problem would arise if we try to classify Bas Van Fraassen's view (1980) that science is not in the business of fact stating. I don't doubt that issues of realism and antirealism are relevant to questions about reductionism and antireductionism. But I do think that we can understand why quasi-realist views fail to fit nicely into the standard framework, for they intend to reject assumptions of that framework that lie even deeper than the common assumptions that presently concern me. Dennett's approach is simply not a position on the dispute between reductionists and antireductionists. So it is no embarrassment that it does not fit neatly into my examination of that debate.

4. Metaphysical Anxieties and the Autonomy Thesis

I began by suggesting that various approaches to reductionism and antireductionism are responses to worries that are metaphysical in one of two ways: they are worries about some metaphysical problems, or they are worries about the nature and legitimacy of metaphysics as an endeavor. It is worth making these anxieties explicit.

As I see it, Metaphysical Reductionists like Kim are driven by the worry that any entities that cannot be reduced will prove to be epiphenomenal or unreal. Similarly, Jackson's approach to reduction begins with the idea that any non-basic entities that cannot be "located" (i.e., reduced) are at best mysterious, certainly epiphenomenal, and probably non-existent. So it is anxiety over having to deny the efficacy or existence of ordinary or important entities (e.g., tables or minds) that drives Kim's and Jackson's reductionism. The Metaphysical Antireductionist has the very same concerns about ordinary or important entities, but just the opposite view of what counts as legitimizing such entities. Antony and Levine are explicit about the anxiety that clearly nagged Fodor, Putnam, and Kitcher, namely that reduction amounts to elimination. To protect tables and minds, we must show that there are generalities about or regularities in the world that essentially involve tables and minds, so that any ontology that leaves them out would be plainly incomplete. Reduction, they say, would result in "losing our minds," so antireductionism is the only way to go.

In contrast, the Antimetaphysical Reductionist and Antimetaphysical Antireductionist are anxious not about particular metaphysical results, but about the endeavor of metaphysics altogether. Usually their motivations are broadly empiricist: Our ontological commitments, if we call them that, should not outrun the evidence that we have. The Antimetaphysical Antireductionists typically accept the official views of Fodor, Putnam, or Kitcher. They observe that as a matter of fact the sciences (and disciplines within sciences) are more or less methodologically independent so that, for example, whether string theory proves to be correct is entirely irrelevant for explanations of evolutionary change. These sciences simply work at different levels. There may sometimes be interesting correlations between the levels, some of which may be explanatorily or methodologically fruitful. But that tells us nothing about the legitimacy of various explanations, nor about their ontological commitments or even whether they have any.

Antimetaphysical Reductionists also shun the ungrounded claims of "speculative" metaphysics, but their anxiety is the antimetaphysical twin of Jackson's Metaphysical Reductionism: To recognize explanations or entities that cannot be located (here, scientifically rather than analytically) among the basic stuff of the universe would be to commit oneself to a metaphysical claim that has no justification and no meaning from the perspective of the sciences. It is true that there are many sciences and many levels of explanation, and these may be methodologically or heuristically fruitful. But claims that cannot be reduced must be regarded as approximations at best, and at worst as simply false.

I have little trouble understanding how epiphenomenalism, eliminativism, meaninglessness, and falsehood can generate anxiety. If reductionism or antireductionism had those consequences, then I would be troubled as well. But I think they do not, and thus that antireductionism and reductionism are solutions to problems that we do not have. It seems to me that these concerns stem from a common but problematic assumption.

Consider the view about reduction and autonomy discussed earlier with respect to Metaphysical Antireductionism. Take as our starting point Antony and Levine's idea that "a property is real (or autonomous) just in case it is *essentially* invoked in the characterization of a regularity" (Antony & Levine, 1997, p. 91). I call this the *autonomy thesis.* Let's clean it up a bit:

> Autonomy: a property x is real if and only if x is essentially involved in (the explanation of) a regularity G.

I'm not going to fuss too much about the exact formulation of the autonomy thesis, as long as we have the general idea that we will ontologically or epistemically commit ourselves to x if and only if there is some regularity Gx that cannot be reduced to any G*y such that $(y \neq x)$.[12] Autonomy says this, I believe, because I interpret the "essentially invoked in" clause in terms of irreducibility. Plausibly this irreducibility has as its domain any reductionist thesis rather than some particular reductionist thesis. So autonomy cuts across the in-house disputes over particular reductionist theses and relations R. According to me, both reductionist and antireductionist approaches, be they metaphysical or antimetaphysical, accept the autonomy thesis and *ipso facto* make an assumption about the relation between explanations and ontological commitment.

The autonomy thesis is a broadly Ockhamist principle, for it asserts that there is only one real ontology and one true story of the world. This follows from the fact that autonomy only recognizes those entities that appear necessarily in the explanation of a regularity. If one could invoke an alternate explanation that does not appeal to some entity, including some regularity, then the entity is not recognized as real. Once we eliminate all the unnecessary and unreal entities, we will be left with the one true explanation and thus the one real ontology. Reduction is compulsory when it is possible: If some entity or explanation can be reduced then it must be reduced, for to recognize a reducible entity or explanation would be to allow redundant entities and explanations. Redundant entities and explanations do no work ontologically or explanatorily, they are at best epiphenomenal or approximate.

Reductionists and antireductionists disagree over whether, for example, pains or genes are reducible or whether they are essentially (irreducibly) involved in (the explanation of) some regularity G. Metaphysical Reductionists and Metaphysical Antireductionists agree, however, that whatever is real must be essentially involved in (the explanation of) some regularity. They both accept the autonomy thesis, and they interpret it as a constraint on ontological commitment in general. Thus the reductionist aims to preserve the reality of pains and genes by showing that they can be located among the basic constituents of reality. (For if they cannot be so located, then they are unnecessary and thus unreal.) And the antireductionist aims to preserve the reality of pains and genes by arguing that our ontology must be expanded to include pains and genes on their own. (For if they do not appear in their own right, then they will be demoted or eliminated in favor of the reduced ontology.)

The Antimetaphysical Reductionist also accepts the autonomy thesis, but interprets it in its antimetaphysical guise. The Churchlands and Bickle argue that scientific practice in fact reduces its ontology and explanations to the lowest available level whenever possible. Of course both the ontology and explanatory value are counted only internally to the practices of a scientific community. Likewise, Antimetaphysical Reductionists hold that a version of the autonomy thesis is part of and internal to scientific practice rather than a general constraint on ontology, for they eschew metaphysical claims about ontology altogether. If the autonomy thesis were interpreted metaphysically then it would make a claim that is external to scientific practice and therefore undecidable or meaningless. The antimetaphysical approach is metascientific. It appeals only to the practices of scientists, and not to any general ontological principles like Ockham's Razor or Alexander's Dictum. The Antimetaphysical Reductionist approach is doubly a response to the autonomy thesis: It rejects the metaphysical interpretations endorsed by the Metaphysical approaches on the grounds that external ontological claims are baseless or meaningless. And it accepts the internal interpretation of the autonomy thesis according to which it is simply a description of the actual practices of some scientists. There is no doubt that the view is radical, and that there is room to question the empirical claims (Looren de Jong & Schouten, 2005; Bickle, 2005, Aizawa, forthcoming). But it is clear how one could subscribe to this approach and what its appeal is meant to be.

Matters are a bit more complicated for the Antimetaphysical Antireductionist. This was the official doctrine of Fodor, Putnam, and Kitcher, recall. According to this view, the legitimacy of higher-level "special science" explanations and their ontologies does not require their reduction. But on the official view, while the special sciences do not require reduction, they are nevertheless compatible with reduction. And this looks like a rejection of the autonomy thesis. If so, then I am mistaken that all four approaches endorse the autonomy thesis. But recall that the Antimetaphysical Antireductionists, while attempting to marginalize reduction, still views reduction as a threat to the legitimacy of higher-level sciences and explanations. This much is clear from the very fact that Putnam, Fodor, and Kitcher all flirt with the stronger position according to which reduction is not just irrelevant but moreover impossible. And Fodor, for example, is explicit that psychological kinds are functional, irreducible, and autonomous (1998).[13]

If I am right, then all four of the above approaches to reductionism can be understood as ways of responding to the perceived consequences of the autonomy thesis, of saving precious ontological commitments or of avoiding the entire ontological business. But what can be said in favor of the autonomy thesis? Why should we suppose that an object exists only if it figures essentially in an explanation? And why should we suppose that an explanation is legitimate only if it is compulsory? Of course two explanations may come into conflict in any number of familiar ways. But it is open to us to reject the idea that two explanations must come into conflict merely because they are two.

This may seem like the kind of metaphysical speculation that the Antimetaphysical Reductionists spurn. But even their internal claim that the autonomy thesis is a descriptive fact of some sciences can be drawn into question. For we may wonder whether the metascientist can give us independent grounds for individuating sciences and their practices, so that we can usefully answer the question of whether autonomy is accepted in a given science or practice. Clearly, given antimetaphysical commitments, there can be no appeal to any an external perspective from which to individuate sciences. But, on the one hand, if we cannot have independent grounds for individuation then the claim that some science accepts autonomy and is reductive in its practices will appear ad hoc. And on the other hand, it seems that the only way for an independent individuation to be is to take up the non-authoritative perspective of another science, thus beginning a regress for the question of how to count sciences and their practices. This dilemma is the practical face of the philosophical puzzle of demarcating internal and external perspectives.

Needless to say, to question the autonomy thesis is not to show that it is false. For present purposes we can settle for the recognition that the four approaches to reductionism are responses to anxieties about the consequences of the autonomy thesis. That recognition allows us to take seriously the possibility that we could tame our anxieties by adopting one of the four responses, or by rejecting the autonomy thesis. I prefer the latter route, and I believe that arguments against the autonomy thesis can be given (Polger, 2004). The dispute over reductionism becomes secondary once the autonomy thesis is rejected. Still it is useful to consider what the terrain looks like, to borrow an expression, after reductionism.

5. Pluralism and Naturalism

To reject the autonomy thesis is to reject the idea that there is only one true explanation or theory whose ontological commitments tell us all that there is. It is to adopt a form of pluralism. The Antimetaphysical Antireductionists hoped to advance a view like this, but because they accepted the autonomy thesis they had to fall into Metaphysical Antireductionism. The pluralistic view is genuinely nonreductive – that is, neither reductive not antireductive. Whether reductionism is true is simply not the right question to ask. So the view that I advocate might be called nonreductive pluralism, and it might just as well (though more awkwardly) be called nonantireductive pluralism. I call it pluralism, naturalistic pluralism or, in some moods, simply naturalism.

Pluralism should not be a free lunch. There are many questions that a pluralist must answer and I am not going to offer you those answers herein. Some of the problems for pluralism are successors to the puzzles that used to face reductionism or antireductionism. But I am optimistic that we can make better progress on them once we leave behind the constraints imposed by the reductionist/antireductionist framework. Only time will tell whether that optimism can be borne out. But it is worth mentioning a few salient features of pluralism as I understand it.

First, pluralism, as I have structured my discussion, is an approach rather than a thesis. There are many particular pluralist theses, and a responsible pluralist will have to figure out how to decide among them. Here I have proposed only the approach, and not any specific pluralistic thesis or theory. If my reasoning is at all compelling, then I may hope to restructure the debate but not to settle it.

Second, pluralism, as I use the label, is simply the rejection of the autonomy thesis: it is the view that there may be more than one explanation for a phenomenon, and that ontology may include items that are not mentioned essentially in a compulsory explanation. Pluralism, therefore, does not guarantee that there is more than one explanation for a phenomenon, or that there is even one. And it certainly does not claim that all explanations are correct or otherwise acceptable. It is an open question to what extent this brand of pluralism (i.e., denial of the autonomy thesis) resembles other forms of pluralism, and whether for better or for worse. Still, it is clear enough that the present thesis deserves the label.

Third, pluralism is not dualism. Dualism is the view that there are two kinds of stuffs or two kinds of properties, the physical and the mental or the inorganic and the organic, for example. According to dualism, these two kinds of stuff or properties are such that one cannot be used to explain or understand the other. But it is not part of pluralism that life or minds cannot be explained in terms of inorganic physical properties or entities. Nor is pluralism a return to the ancient view that there are more than one or two basic kinds of stuff that compose everything else. Pluralism is not a view about how many basic kinds of things or properties there are, or even about whether there are any basic things or properties. It is merely the denial of the autonomy thesis.

Finally, as I have already noted, pluralism does not make every metaphysical or explanatory question go away. It simply resituates them. For example, I take it that explanatory exclusion problems get no traction for an explanatory pluralist, more or less by definition. On the other hand, pluralism splinters the question of which explanation to prefer into equally pressing questions about the relations among explanations and about when explanations come into genuine conflict. And while explanatory exclusion is no problem, it seems like causal exclusion concerns simply get a new formulation but otherwise look much as they always did. Understanding the relations that hold among plural explanations looks to be at least as troublesome as finding some unified reduction relation. If it is too much trouble then we shall have to think about abandoning pluralism for a still better alternative. I claim only we need not restrict ourselves to the false choice between reductionism and antireductionism.

Acknowledgments

The approach sketched in this chapter is an elaboration of the view I defended in 2004, and it is a response my own meta-anxieties that arise from the competing tugs of all the first-order anxieties. I am not sure that I have found the cure, but I sometimes think that I am making progress. I would like to thank the following therapists for their assistance on this matter: Ken Aizawa, Marica Bernstein, John Bickle, Robert Brandon, Carl Craver, Owen Flanagan, Carl Gillett, Greg Johnson, Larry Jost, Tony Landreth, Michael Lynch, Bob Richardson, Larry Shapiro, Rob Skipper, and the editors and reviewers for this volume.

Notes

1 See, e.g., Kitcher (1984), Schaffner (1993).
2 It's not obvious that it makes sense to hold the theoretical thesis while denying the ontological thesis. At any rate I cannot think of any examples.
3 But not always. For example, if, as in John Bickle's (2003, and this volume) "New Wave" reductionism, reductionism should be understood as a descriptive claim about the actual practices of a certain group of scientists, then the choice of reductive target (and the group of scientists) will be absolutely crucial. If Bickle is right then reductionism is not a general doctrine after all and we have no reason to expect various reductionisms to have much in common.
4 This quotation reveals that Bickle now views his earlier (1996, 1998) "new wave reductionism" as having been, paradoxically, a version of what I am calling Antimetaphysical Antireductionism. Perhaps this recognition explains why he has shifted to a more "ruthlessly" reductive "new wave metascience" (2003).
5 A fuller discussion of this view can be found in Polger (2004, ch. 6).
6 It seems fair to say that reduction is mainly an issue among realists, but see Section 2.5, below.
7 I do not mean to suggest that there are not similar problems for the other approaches. I'm merely trying to make clear, in this case, how an Antimetaphysical Antireductionist might be attracted to Metaphysical Antireductionism in the way that I suggest below.
8 For the record, the reductive part of the Antony and Levine "reduction with autonomy" view seems to involve an endorsement of the Metaphysical Reductionist approach followed by Kim, which worries that irreducible states or properties will prove unreal due to causal exclusion and Alexander's dictum.
9 And Chalmers himself acknowledges that facts about causation will be exceptions to the reductive picture that he paints for the physicalist, so causal generalizations will not be reducible to non-causal physical facts (1996).

10 There are senses in which one might think of Block and Stalnaker as reductionists. But the dispute between them, on the one hand, and Chalmers and Jackson (2001), on the other hand, is over which relations yield the demanding kind of reduction that Chalmers and Jackson suppose that physicalism requires. Block and Stalnaker deny that the Chalmers and Jackson style reduction obtains. Thanks to an anonymous referee for this chapter for reminding me of this point.

11 Bickle's (1998) model of reduction is similar to that suggested by Schaffner (1967).

12 Obvious the notion of a "regularity" will be ambiguous between a real pattern in the world and a statement characterizing such a pattern, in the same way that the notion of "law" is ambiguous. And context will normally disambiguate the interpretation of "regularity" just as it does for "law."

References

Aizawa, K. (forthcoming). The biochemistry of memory consolidation: A model system for the philosophy of mind. *Synthese.*

Antony, L., & Levine, J. (1997). Reduction with autonomy. In J. Tomberlin (Ed.), *Philosophical Perspectives 11: Mind, Causation, and World* (pp. 83–105). Boston: Blackwell.

Bickle, J. (1996). New wave psychophysical reductionism and the methodological caveats. *Philosophy and Phenomenological Research, 56(1)*, 57–78.

Bickle, J. (1998). *Psychoneural Reduction: The New Wave.* Cambridge, MA: MIT Press.

Bickle, J. (2003). *Philosophy and Neuroscience: A Ruthlessly Reductive Account.* Dordrecht: Kluwer Academic Publishers.

Bickle, J. (2005). Molecular neuroscience to my rescue (again): Reply to Looren de Jong & Schouten. *Philosophical Psychology, 18*, 487–494.

Block, N., & Stalnaker, R. (1999). Conceptual analysis, dualism, and the explanatory gap. *Philosophical Review, 108*, 1–46.

Carnap, R. (1950/1956). Empiricism, semantics, and ontology. *Revue Internationale de Philosophie, 4*, 20–40. Reprinted in *Meaning and Necessity: A Study in Semantics and Modal Logic.* Chicago: University of Chicago Press.

Chalmers, D. (1996). *The Conscious Mind: In Search of a Fundamental Theory.* New York: Oxford University Press.

Chalmers, D., & Jackson, F. (2001). Conceptual analysis and reductive explanation. *Philosophical Review, 110*, 315–361.

Churchland, P. M. (1981). Eliminative materialism and the propositional attitudes. *Journal of Philosophy, 78*, 67–90.

Churchland, P. M. (1982). Is "thinker" a natural kind? *Dialogue*, *21*, 223–238.

Churchland, P. S. (1983). Consciousness: The transmutation of a concept. *Pacific Philosophical Quarterly*, *64*, 80–93.

Dennett, D. (1978). Intentional systems. In D. C. Dennett, *Brainstorms* (pp. 3–22). Hassocks: Harvester.

Dennett, D. (1987). *The Intentional Stance.* Cambridge, MA: MIT Press.

Fodor, J. (1974). Special sciences (or the disunity of science as a working hypothesis. *Synthese*, *28*, 97–115. Reprinted in N. Block (Ed.), *Readings in Philosophy of Psychology* (Vol. 1, pp. 120–133). Cambridge, MA: Harvard University Press, 1980.

Fodor, J. (1998). Special sciences: Still autonomous after all these years. In J. Tomberlin (Ed.), *Philosophical Perspectives 11: Mind, Causation, and World* (pp. 149–163). Malden: Blackwell.

Jackson, F. (1982). Epiphenomenal qualia. *Philosophical Quarterly*, *32*, 127 136.

Jackson, F. (1998). *From Metaphysics to Ethics: A Defense of Conceptual Analysis.* Oxford: Oxford University Press.

Kim, J. (1993). *Supervenience and Mind: Selected Philosophical Essays.* New York: Cambridge University Press.

Kim, J. (1998). *Mind in a Physical World: An Essay on the Mind–Body Problem and Mental Causation.* Cambridge, MA: MIT Press.

Kim, J. (2005). *Physicalism, Or Something Near Enough.* Princeton, NJ: Princeton University Press.

Kitcher, P. (1984). 1953 and all that: A tale of two sciences. *Philosophical Review*, *93*, 335–373.

Lewis, D. (1994). Lewis, David: Reduction of mind. In S. Guttenplan (Ed.), *A Companion to the Philosophy of Mind.* Oxford: Blackwell.

Looren de Jong, H., & Schouten, M. K. D. (2005). Ruthless reductionism: A review essay of John Bickle's *Philosophy and Neuroscience: A Ruthlessly Reductive Account. Philosophical Psychology*, *18*, 473–486.

Oppenheim, P., & Putnam, H. (1958). Unity of science as a working hypothesis. *Minnesota Studies in the Philosophy of Science*, *2*, 3–36.

Polger, T. (2004). *Natural Minds.* Cambridge, MA: MIT Press.

Putnam, H. (1975). Philosophy and our mental life. In H. Putnam, *Mind, Language and Reality: Philosophical Papers* (Vol. 2). New York: Cambridge University Press.

Schaffner, K. (1967). Approaches to reduction. *Philosophy of Science*, *34*, 137–147.

Schaffner, K. (1993). *Discovery and Explanation in Biology and Medicine.* Chicago: University of Chicago Press.

van Fraassen, B. (1980). *The Scientific Image.* Oxford: Oxford University Press.

4

THE METAPHYSICS OF MECHANISMS AND THE CHALLENGE OF THE NEW REDUCTIONISM

Carl Gillett

1. Introduction

Over the twentieth century, as Figure 4.1 graphically illustrates, scientific investigations have given us a detailed account of many natural phenomena, from molecules to manic depression, through so-called "mechanistic," or "functional," explanations based upon our understanding of the entities that compose these phenomena.[1] Given their importance, there has unsurprisingly been a long tradition of philosophical research on mechanistic explanations.[2] And this work has played an especially important role in debates over reduction, since nonreductivists, such as Hilary Putnam, Jerry Fodor, William Wimsatt, and Philip Kitcher, have all used detailed scientific evidence about the compositional relations posited between properties in mechanistic explanations, or so-called "realization" relations, to show that the Nagelian model of reduction (Nagel, 1961), and its semantic descendants, all face grave problems (see Putnam, 1975a, 1975b; Fodor, 1974, 1975; Wimsatt, 1974, 1976; Kitcher, 1984, amongst others). In various ways, these writers have all used the widespread phenomenon of multiple realization to show that we cannot plausibly establish the Nagelian's required bridge laws or identity statements, and that we also have good reasons to believe that special science predicates are indispensable, contrary to the claims of such reductive positions.

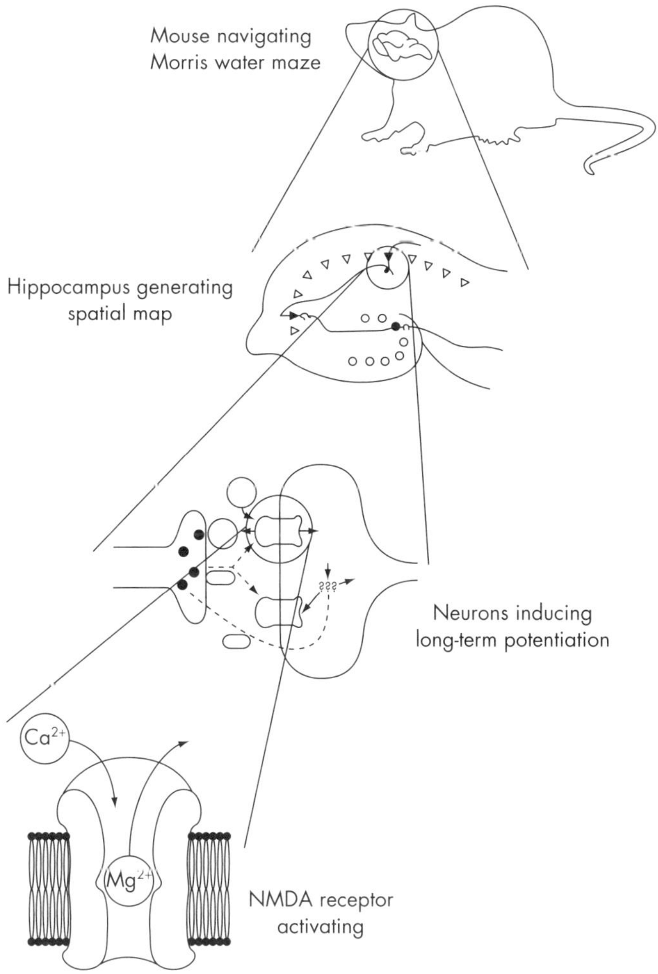

Figure 4.1 Levels in the hierarchical organization of the mechanism of spatial memory. *Source*: Carl F. Craver and Lindley Darden, "Discovering Mechanisms in Neurobiology: The Case of Spatial Memory," in Peter K. Machamer, Rick Grush, and Peter McLaughlin (eds), *Theory and Method in the Neurosciences*. © 2001 by University of Pittsburgh Press. Reprinted by permission of the University of Pittsburgh Press.

Rather oddly, despite their important role in debates over reduction, until very recently there has been little detailed work on the nature of the compositional relations posited in mechanistic explanations. The Positivists' influence appears to be largely behind this relative neglect, for their famous proscription of metaphysics has meant that philosophers of science have been leery of addressing topics of an explicitly metaphysical caste. However, the pioneering work of Jaegwon Kim has inspired a flowering of interest in scientific composition, including realization relations.[3] And, simultaneously, we have also seen a recent resurgence of interest in the "metaphysics of science" more generally.[4] The resulting metaphysical work is not the *a priori* inquiry rightly disparaged by the Positivists, but is instead the careful, abstract investigation of ontological issues as they arise within the sciences and their explanations, findings, models, and so on.

Work on the metaphysics of science has the potential to transform many important debates in the philosophy of science and Kim (1998, 1999) has suggested that this is very much the case with discussions of reduction. He has argued that if reductionists embrace both the metaphysics of science and the nonreductivists' insights about composition in the sciences, then evidence about realization, and other compositional relations, is transformed into the engine of a new, ontological form of reductionism. My goal in this chapter is to assess Kim's claims. I will therefore provide a deeper characterization of the metaphysics of mechanisms and the compositional relations they involve. And I shall then use this framework to show that understanding scientific composition does indeed drive a novel and prima facie plausible form of ontological reduction as Kim predicts, though one that is rather different from, and far more radical than, Kim's own formulation of it.

To this end, I will continue recent work illuminating the nature of scientific composition, in Section 2, by providing an outline of a general metaphysical framework for mechanisms in the sciences. As we will see, such mechanisms involve an array of integrated compositional relations between powers, properties, individuals, and the mechanisms that they ground. I will provide a general overview of the features of such compositional relations, highlighting how they drive mechanistic explanations, and also a precise definition of the realization relations between properties. I will then use this framework, in Section 3, to illuminate a form of ontological reduction often espoused by working scientists, which I will call "Compositional" reduction. This reductionism embraces the compositional relations posited in ongoing scientific research and uses the true statements illuminated by such concepts. For the Compositional reductionist argues that, upon more careful reflection, the very nature of these posited relations entails we should ultimately take the fundamental component entities to be the *sole* truthmakers for such special science statements. Ultimately, the Compositional reductionist thus concludes that concepts of composition in the sciences are just that: useful concepts for illuminating true statements in ongoing scientific research, but concepts which ultimately do not literally reflect the structure of the natural world. I will show in detail how such a reductionism proceeds for special science properties, but also sketch how it applies to special science powers, individuals, and processes as well.

Initially, Compositional reduction appears puzzling, even incoherent, to many philosophers of science, since it reverses long-standing commitments in debates over reduction in the sciences. Perhaps most strikingly, Compositional reductionism combines a trenchant ontological reductionism with a comprehensive nonreductivism about special science predicates, explanations, and theories. In order to better illuminate the nature of Compositional reduction I will therefore examine, in Section 4, how it deals with the two main arguments for "nonreductivism" based around, respectively, the phenomenon of multiple realization and the, in principle, indispensability of special science predicates. I will show that Compositional reductionism is untouched by both objections, for this position embraces the key insights of such critiques while still pressing its ontological reductionism. I will consequently suggest that Compositional reductionism radically reorients contemporary discussions of reduction.

I should carefully note that I do not endorse Compositional reductionism and that I have critically examined it at length elsewhere.[5] Nonetheless, my final conclusion here will be that, as Kim predicts, once we have a more adequate framework for the concepts of composition used in the sciences, then we can finally illuminate a powerful brand of reductionism that has long attracted working scientists. For we clearly see how, and why, the realization, and other compositional, relations illuminated by mechanistic explanations provide the engine of prima facie plausible version of ontological reductionism, though one that embraces the legitimacy, and in principle indispensability, of special sciences

2. The Metaphysics of Science and the Metaphysics of Mechanisms

The neurosciences are a very useful case for our purposes because they quickly show how comprehensive mechanistic explanations, and the compositional relations they posit, actually are in the sciences. The general type of scientific example with which we are concerned is again nicely framed by Figure 4.1. For, as the diagram highlights, we now have good evidence that even psychological states, like spatial memory, are plausibly composed by brain areas; where these brain areas are themselves composed by collections of cells like neurons; where neurons in turn are composed of cellular structures, like ion channels; furthermore, though not shown in our diagram, ion channels are also plausibly composed by complex biochemical molecules or "protein sub-units"; where such protein sub-units are themselves composed by atoms; where atoms are composed by the sub-atomic entities studied by physics. We thus have a comprehensive hierarchy of composing, and composed, individuals in this caseand these relations between individuals are accompanied by compositional hierarchies for their powers, properties, and processes. Furthermore, similar cases, all centered around the compositional relations posited by mechanistic explanations, are found across the full spectrum of scientific disciplines from materials science to meteorology.

As we will shortly see, such cases of composition in the sciences obviously involve relations between "packages" of powers, properties, individuals, and mechanisms in higher- and lower-level sciences.[6] However, to make our discussion manageable, and because they have been the primary focus of recent debates, I will concentrate primarily upon giving a detailed metaphysical account of the compositional relations between properties and their powers. We thus need a metaphysical framework for properties and, following recent work in the metaphysics of science, I will use a "causal theory of properties" under which a property is individuated by the causal powers it *potentially* contributes to the individuals in which it is instantiated and where I will simply assume all instances of a property contribute the same powers, under the same conditions, in the actual world.[7] For reasons that will become apparent below, I will be concerned with properties I term "causally efficacious," that is, properties whose instantiation *actually* determines the contribution of causal powers to an individual. (In addition, I will follow the metaphysician's use of the term "entity" as a catch-all referring to powers, properties, individuals, and processes, while reserving the term "individual" to refer to individuals).

The basic features of compositional relations can obviously be illuminated by all manner of scientific examples, but for brevity I will

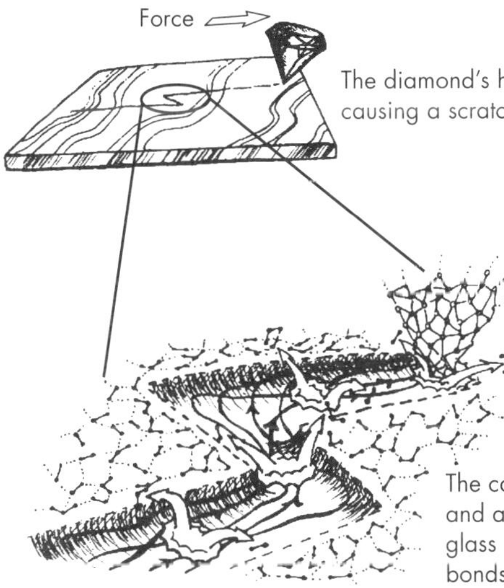

Figure 4.2 The lower-level mechanisms of carbon atoms breaking the bonds between, and displacing, specific glass molecules implement the process of a diamond scratching a pane of glass

consider a case whose details are compact, and widely known, in the extreme hardness of a diamond and the bonding and alignment of carbon atoms. Our cartoon in Figure 4.2 highlights some of the mechanisms involved with this case. When a cut diamond is pressed with force across a pane of glass, in a certain sequence of directions, then the diamond's hardness causes a Z-shaped scratch in the glass. Here we have a causal process, i.e., mechanism, which is grounded by the triggering of one of the diamond's powers, contributed to the diamond by its property of being very hard, which results in a certain effect – a Z-shaped scratch on the glass.

As our cartoon also roughly captures, the sciences have provided an elegant, and familiar, explanation of this higher-level process by illuminating its implementation. For the sciences have shown how the lower-level mechanisms of particular carbon atoms breaking the bonds between, and displacing, specific glass molecules compose, or more specifically "implement," the process of the diamond scratching the glass. For, under the relevant background and triggering conditions, the powers of the carbon atoms, to hold each other in close relative spatial relations and to break the bonds between glass molecules when under pressure, are triggered in a host of carbon atoms. The result is a large number of causal processes involving particular carbon atoms, each of which cause specific glass molecules to break their bonds with other glass molecules and change their spatial locations. It is worth marking that none of these lower-level mechanisms involving carbon atoms results in an effect that has a macroscopic breadth and depth, or has a Z-shape, or any of the other features of the process of the diamond scratching the glass. Instead, when triggered, the powers of particular carbon atoms each ground a mechanism that causes certain glass molecules to break their bonds and change position. Nonetheless, the many mechanisms involving carbon atoms *together* implement, i.e., non-causally result in, the qualitatively different process involving the diamond.

Against this backdrop, since properties are our primary focus, let us more carefully consider the compositional relations posited in this type of case between property instances and their powers, starting with the latter. The sciences arguably show that the powers of carbon atoms compose, or as I shall put it "comprise," the powers of the diamond. Why? To start, we must mark that a power basically is an entity whose possession by an individual allows this individual to enter into certain causal processes. Powers are thus plausibly individuated by the particular mechanisms which their triggering grounds. But we have just seen that the sciences show that, if triggered, the powers of the carbon atoms ground mechanisms that together implement the mechanism grounded by the powers of the diamond, if they are triggered. Thus we can plausibly see that the sciences illuminate how, and why, the powers of the carbon atoms *together* comprise, i.e., non-casually result in, the powers of the diamond, since the mechanisms grounded by the powers of the carbon atoms implement the mechanisms grounded by, and individuative of, the diamond's powers.

Similar points hold for the compositional relations posited between the property instances in such examples. A property instance is an entity that makes a difference to the causal powers of individuals, usually by contributing powers to individuals. Most scientific properties are thus individuated by the causal powers they potentially contribute to individuals under certain background conditions. We can consequently see that the sciences again illuminate how, and why, the properties and relations of the carbon atoms compose, or "realize," the properties of the diamond. For we have now seen that the powers of the carbon atoms comprise, i.e., non-causally result in, the powers individuative of the properties of the diamond such as its property of being very hard. As a result, the sciences explain why if one has the carbon atoms with their properties and relations, then the latter properties will *together* realize the diamond's property of hardness. For the carbon atom's relations of bonding and alignment together contribute powers that comprise the powers individuative of the property of being very hard and hence non-causally result in, and realize, an instance of the property of hardness in the diamond.

In our example, we therefore have compositional relations between mechanisms, properties, and powers. Furthermore, we plausibly also have composition between individuals, since the sciences show the carbon atoms are *parts* of the diamond and hence compose it too (or "constitute" it in my terms). In the face of such complex and multiple relations, some philosophers of science have worried that these matters cannot be made precise.[8] We can assuage such concerns by offering a more detailed "thumbnail" account of the interconnected natures of the comprising of powers, and realization of properties, which will also be useful for our later work on reduction. For example, we may roughly frame comprising thus:

> (Comprising) Powers C1-Cn, had by individuals *s*1-*s*n (or individual *s*), comprise the power C*, had by individual *s* under background condition \$, *if and only if*, the mechanisms grounded by the triggering of powers C1-Cn, under triggering conditions \$t1-\$tn and background condition \$, would together implement the causal mechanisms grounded by the triggering of C*, under triggering conditions \$t1-\$tn and background condition \$, but not vice versa.

Building upon this account, we can also frame a view of the realization of properties, in a version of the so-called "Dimensioned" account, as follows:

> (Realization) Property instances F1-Fn, in individuals *s*1-*s*n (or individual *s*), realize a property instance G in an individual *s*, under background conditions \$, *if and only if* the powers individuative of G, in *s* under \$, are comprised by the powers contributed by F1-Fn to *s* or *s*1-*s*n, which are constituents/parts of *s*, but not vice versa.

Obviously, this "thumbnail" sketch of comprising and realizing needs to be supplemented by accounts of both the implementation of mechanisms and the constitution of individuals. Elsewhere I have offered a detailed, and fully integrated, account of the compositional relations between powers, properties, individuals, and processes, but for my purposes here our simple "thumbnail" definition will suffice.[9]

Following the work of the handful of seminal writers, such as Donald Campbell, Herbert Simon, and William Wimsatt, who earlier sought to frame what is common to compositional relations in the sciences, we can now also provide a more general account to complement such detailed definitions (see Campbell, 1958; Simon, 1969; Wimsatt, 1974, 1976, 1994). For we have found that compositional relations all share a range of features. First, we should mark that such relations are obviously a species of determination relation, but are rather different from causal relations. Compositional relations are not temporal in nature, since their "upward" determination is instantaneous, do not relate wholly distinct entities, and do not involve the transfer of energy and/or the mediation of force. In contrast, the "horizontal" determination involved with causation is temporally extended, does relate wholly distinct entities and often involves the transfer of energy and/or the mediation of force. Composition is thus a variety of what I have been terming "non-causal" determination.

Second, scientific composition usually relates qualitatively different kinds of entity. For example, diamonds have properties like hardness, and hardness contributes the powers to scratch glass. But no carbon atom has the property of hardness, nor any property that contributes the power to scratch glass. In our case, we thus have individuals that constitute other individuals with which they share no properties, properties that realize other properties with which they share no common powers, and similar points hold for the relevant powers and mechanisms. A quick examination of the compositional relations illuminated by the full spectrum of special sciences shows that this feature is common to all of their findings. (For example, think about the features of the entities involved in the compositional relations outlined in Figure 4.1.)

Third, and perhaps most importantly, we should also mark a common feature of scientific composition that apparently illuminates how the sciences can use compositional relations to provide such distinctive, and powerful, explanations of one kind of entity in terms of qualitatively different kinds of entity. For, given their interconnections, the entities studied by lower-level sciences *together* compose the qualitatively different entities studied by higher-level sciences. The simple secret of compositional relations, and the mechanistic explanations that posit them, is therefore that, although individually the component entities are qualitatively different from the composed entity, nonetheless the components *together* non-causally result in the composed entity. This distinctive feature of compositional relations consequently allows one to mechanistically explain entities of one kind in terms of entities of very different kinds. And, since it will be useful in my later discussion, it is worth highlighting how this characteristic also underlies the phenomena that so much of the recent debate over reduction has focused upon.

Putting things crudely, it is because realizer properties together realize some special science property that we get multiple realization – for distinct combinations of lower-level properties, that are very different from each other, may nonetheless each still together non-causally result in instances of the same special science property. Furthermore, this point also shows why we rarely get bridge laws or identities in the sciences. Earlier writers, like Putnam, Fodor, and Kitcher, highlighted the fact that the nature of scientific composition means we usually have multiple composition bases and rarely get exact coextension between any particular combination of components and a composed entity. Thus earlier writer persuasively argued that Leibniz's Law is usually violated for any particular combination of components and a specific composed entity and we thus usually fail to get identities. Similar points hold with regard to bridge laws. Furthermore, our more detailed overview of scientific composition highlights still further problems precluding identities. Compositional relations are asymmetric, many–one (i.e., they relate many components and one composed entity) and relate entities of qualitatively different kinds. In contrast, and for obvious reasons, identity is always symmetric, one–one and relates entities of exactly the same kind. We can therefore see that, given their differing natures, compositional relations in the sciences usually will not ground identities between the higher and lower-level entities that they relate.[10]

As I have noted at various points, much more work is needed in order to provide a full account of scientific composition. For example, I have said nothing about whether the foregoing framework simply captures the nature of the *concepts* of composition used by working scientists in the course of their ongoing research, or whether our framework describes the character of actual compositional *relations* holding between entities in the world. I have deliberately left this question to one side, for in the next section I want to apply our work on scientific composition to reconsider the nature of "reduction" in the sciences. And I will highlight a novel form of reduction based upon the contention that the framework of this section only illuminates the *useful concepts* used by working scientists, for this reductionism argues that we should ultimately not accept the existence of any composed entity, and hence any relation of composition, in nature itself.

3. Illuminating Ontological Reductionism

Many working scientists are attracted to ontologically reductive views, for example Weinberg (1992), arguing that the special science entities mechanistically explained by being shown to be composed by the entities of lower-level sciences are "nothing but," or "nothing more than," or "nothing over and above" the lower-level entities that have been shown to compose them. With our better grip on the overall features of the compositional relations posited in scientific research we can now understand the pull of such intuitions that the appearance of distinct entities bearing relations of compositions is just that – an *appearance*. The key is to follow Kim's radical advice that reductionists finally throw off any lingering prejudices, derived from the Positivists, in order to pursue the metaphysics of science, and a thoroughly ontological approach, while also embracing the nonreductivists' insights about composition in the sciences. As a result, the reductionist is then in a position to argue that, given the very nature of concepts of "composition" used in scientific research, we can account for the truth of the relevant special science statements previously thought to refer to composed entities simply by taking components to be the sole truthmakers. The broached reductive reasoning eschews both bridge laws and identities, for it is instead a simple type of ontological parsimony argument driven by the nature of the concepts of compositional relations used in ongoing scientific research. To illuminate this position I will now provide a detailed version of such an ontologically reductive argument for special science property instances. (Later I will show that similar reasoning can be provided for special science powers, individuals, and mechanisms.)

The argument of this ontological reductionist about property instances proceeds as follows. Given the putative nature of realization, the reductionist may argue it is ontologically profligate to take any realized property instance to determine the contribution of causal powers, and hence to be causally efficacious, *in addition* to its realizer property instances. For the reductionist notes that, given the nature of the realization relation, we can account for *all* the causal powers of individuals simply using the contributions of powers by the realizer property instances of these individuals, or their constituents, rather than also as contributions from realized property instances. But we cannot account for all causal powers of individuals simply as contributions by realized property instances. If we assume that the contribution of the causal powers of individuals is not overdetermined, then appealing to Occam's Razor the reductionist argues that we should accept the existence of no more causally efficacious property instances than we need to account for the causal powers of individuals. The proponent of this simple argument thus concludes that we should *only* accept that realizer property instances contribute powers to individuals and hence *only* accept that realizer property instances are causally efficacious.[11] But it is also plausibly true that the only property instances that exist make a difference to the causal powers contributed to individuals. We may thus further conclude that there are only realizer property instances. Let us call this the "Extended Argument from Realization."[12]

It bears emphasis that this argument is not, in the first instance, focused on *causation*, or causal exclusion principles, or concerns about causal overdetermination, but instead argues to a conclusion about the putative implications of the *compositional* relations posited between properties in mechanistic explanations in the sciences. It should also be clear that this is a reductive argument that is thoroughly metaphysical in nature. The Extended Argument from Realization allows us to reduce not just our commitments to efficacious property instances, as arguments focused on "mental causation" attempt, but also to reduce our commitments to realized property instances outright. From this point I will use the terms "ontological" or "Compositional" reductionism to refer only to the brand of reduction that utilizes the Extended Argument from Realization or analogous parsimony arguments using other scientific composition relations.

So far we have provided an explicit ontological reductionism about property instances, and it will be helpful at this point if we briefly return to our general view of composition in the sciences, for we can consequently illuminate the deeper rationale of Compositional reductionism and show that it extends to reduce all special science entities. As we have seen, a general feature of the concepts of composition used in mechanistic explanations is obviously that the nature of the component entities together non-causally, that is, instantaneously and without mediation of force and/or transfer of energy, result in the composed entity. Putting it bluntly, the composing entities together *are* the composed entity. But, upon more careful reflection, the Compositional reductionist argues that, given the nature of these compositional concepts, we can thus see that as truthmakers for any true special science statement about "composition" we should not be committed to *both* the components *and* some further composed entity. For we can consequently account for the truth of the special science statements previously thought to refer to composed entities by simply taking their components to be the sole truthmakers. And if the picture of scientific composition outlined in Section 2 captures the nature of the concepts of the comprising between powers, the constitution or part–whole relations between individuals, and the implementation relations between mechanisms routinely posited by working scientists, then it appears that such a Compositional reductionism also extends to reduce our previous commitments to special science powers, individuals and mechanisms.

At this point it may be useful to contrast Compositional reduction with Kim's own detailed formulation of ontological reduction. We have already noted one divergence, since Compositional reduction explicitly concerns composition, while Kim has often focused upon causal notions or concepts of supervenience. Perhaps more important is the fact that, like the Nagelian, Kim continues to drive his form of ontological reduction using identities – albeit between property instances. But, as we saw in Section 2, the widespread phenomenon of multiple realization makes it plausible that we rarely get such identities. Furthermore, we also saw that a better understanding of the nature of scientific composition illustrates still deeper difficulties precluding identities between special science and lower-level properties and their instances. The unavailability of such identities hobbles Kim's reductionism, but in contrast leaves Compositional reduction untouched since this position has no recourse to anything but concepts of realization and other compositional relations.

To summarize, although diverging from Kim's own formulation of ontological reductionism, our work in this section has vindicated Kim's wider contentions that if the reductionist follows the insights about special sciences rightly trumpeted by nonreductivists, while embracing the metaphysics of science as a tool to unpack these findings, then the result is a prima facie plausible form of global ontological reduction. Furthermore, as we shall see in still more detail in the next section, we have found that Compositional reduction avoids key problems that undermine other versions of reductionism. Using Kim's overarching advice as a guide, we have thus articulated a clear, and relatively precise, argument that illustrates why so many working scientists may be right to assume that through their mechanistic explanations, and the "compositional relations" they posit, the sciences reveal that all special science "entities" are literally "nothing but" the lower-level entities that "compose" them.

4. Indispensable Special Sciences and the Challenge of Compositional Reductionism

Given the recent history of the debates over reductionism, many philosophers of science will likely remain unconvinced of the importance of Compositional reductionism. On the one hand, reductionist philosophers will wonder whether it is even coherent to combine, as Compositional reductions seeks to do, a thorough ontological reductionism with a nonreductivism about special science predicates, explanations, and theories. While, on the other hand, nonreductivist philosophers of science will continue to press the point that actual scientific practice trumps mere metaphysics, suggesting that the phenomena of multiple realization, and evidence about the in principle indispensability of special science predicates, trump Compositional reduction. In response to these concerns, I will examine each of the nonreductivists' challenges in turn, for in answering these objections we will also see how, and why, Compositional reductionism can embrace nonreductivism about special science predicates, explanations, and theories. (In my discussion, for brevity, I will focus solely upon special science predicates.)

In resisting Nagelian reduction, as we saw in Section 2, the common nonreductivist refrain is that the heterogeneity of lower-level realizers (or "kinds"), i.e., the prevalence of multiple realization, precludes "reductionism." One way of framing the driving idea of these critiques, that will again be very useful shortly, is that our world contains many "macro-commonalities," i.e., features shared by aggregations of microphysical individuals, that are "pure" macro-commonalities – that is, features shared by such aggregations that are microphysically heterogeneous. A pure macro-commonality is thus a commonality found at the macro-level without a corresponding commonality at the micro-level. There are obviously "impure" macro-commonalities, i.e., features shared by microphysical aggregations in virtue of their shared microphysical commonalities, but nonreductivists have argued convincingly that the Nagelian's mistake is to assume that all macro-commonalities are of this impure type. As we have seen, nonreductivists have shown that the prevalence of multiple realization, and accompanying pure macro-commonalities, mean that higher- and lower-level entities, and the predicates we use to refer to them, simply do not neatly line up – thus we can rarely formulate bridge laws or identity statements, and the predicates of the special sciences are, in principle, indispensable for capturing such pure macro-commonalities.

However, given its very different nature from Nagelian reduction, Compositional reductionism is untroubled by multiple realization and its implications. We can begin to appreciate this point if we consider a putative special science property E where for distinct instances of E scientists outline mechanistic explanations involving heterogeneous realizer property/relation instances, instantiated in lower-level individuals, which are each such that they suffice for all the powers of E in some higher-level individual. Obviously, such a case is an example of multiple realization because different realizer properties are taken to be responsible, usually in other individuals, for contributing the powers that result in the powers individuative of E. Nonetheless, despite the heterogeneity of these realizers, the Compositional reductionist will argue that these realizers still ground their key argument, since these diverse realizers are all such that together they result in the powers individuative of E. Consequently, using the Extended Argument from Realization the reductionist can argue that, by themselves, the realizers consequently suffice as the truth-makers for the relevant special science statements, and we should not also endorse the existence of any realized instance of the putative special science property E.

The latter point suggests that, far from being undermined by it, Compositional reductionism can embrace statements putatively about multiply realized properties and use them to drive its brand of ontological reduction. However, a couple of obvious questions immediately arise about the stance that Compositional reductionism takes with regard to special science statements and their predicates. Does such a reductionism take special science statements and predicates to be, in principle, dispensable? And if it does not, as I have suggested is the case, then what explanation can the Compositional reductionist give of the utility, and meaning, of such special science statements, and their predicates, if she also endorses ontological reductionism? These are obviously important questions for Compositional reductionism, and in order to address them let us now turn to the second kind of defense of contemporary nonreductivism. To highlight the nature of such arguments based around the, in principle, indispensability of special science predicates I will use work by Philip Kitcher (1984), since it fits well with our earlier discussion. However, I contend that the general approach I outline also works with other recent arguments from writers like Putnam, Fodor, and Wimsatt.

Kitcher takes as his focus the case of molecular biology and various higher biological sciences, such as classical genetics, cytology, and embryology. And Kitcher convincingly argues that properties and individuals putatively referred to by predicates in the higher biological sciences are taken to be multiply realized in the case of properties, and what we might term "multiply constituted" in the case of individuals, by heterogeneous molecular properties and individuals. Furthermore, Kitcher argues at length that attempts to replace key explanations using the predicates of higher-level sciences, whether of classical genetics, cytology, embryology, or others, with explanations using the predicates of molecular biology very often fail. To illustrate the thrust of these arguments for my purposes here it suffices to consider the example that Kitcher uses to summarizes his wider conclusions. For instance, Kitcher tells us:

> The distribution of genes to gametes is to be explained, not by rehearsing the gory details of the reshuffling of the molecules, but through the observation that chromosomes are aligned in pairs just prior to the

> meiotic division; and that one chromosome from each matched pair is transmitted to each gamete. We may formulate this point in the biologists preferred idiom by saying that the assortment of alleles is to be understood at the cytological level. What is meant by this description is that there is a pattern of reasoning . . . which involves predicates that characterize cells and their large scale structures. That pattern of reasoning is to be objectively preferred to the molecular pattern which would be instantiated by the derivation that charts that complicated rearrangements of individual molecules because it can be applied across a range of cases which would look heterogeneous from a molecular perspective. Intuitively, the cytological pattern makes connections that are lost at the molecular level, and is thus to be preferred. (Kitcher, 1984, pp. 22–23)

Here Kitcher presses the indispensability, in framing certain true explanations, of special science predicates referring to what I earlier dubbed "pure" macro-commonalities, i.e., features that are shared between aggregations of microphysical individuals which are molecularly and hence microphysically heterogeneous in their properties and relations. If we solely used the predicates of molecular biology (or microphysics), then Kitcher argues that we would fail to refer to the macro-commonality shared between molecularly and microphysically heterogeneous aggregations that is central in explaining the shared behavior of these aggregations as meiotic processes. As a result, special science predicates referring to the relevant pure macro-commonality are, in principle, indispensable in formulating this and many other true explanations. Kitcher summarizes the underlying basis of his argument by referring to what he terms the "organization of nature" which he suggests entails that special science predicates are indispensable in the ways he outlines, for our scientific evidence shows, putting it in my terms, that when matter aggregates we often have pure macro-commonalities which predicates of lower-level sciences fail to capture (Kitcher, 1984, p. 22). And this point allows us to see how, and why, our Compositional reductionist will respond to arguments like Kitcher's.

If we focus on the entities of physics as the relevant "components," the Compositional reductionist will obviously accept that the entities of physics aggregate, where this is often into vast aggregations involving huge numbers of microphysical individuals with all kinds of microphysical properties and bearing a range of microphysical relations to each other. Furthermore, as we have seen, this ontological reductionist will also take such aggregations of microphysical individuals to share pure macro-commonalities, since the Compositional reductionist accepts that many special science statements refer to pure macro-commonalities, including those statements usually taken to refer to multiply realized properties. Crucially, however, the Compositional reductionist will deny that we should reify such aggregations of microphysical individuals or their features, in pure macro-commonalities, into further individuals or properties. Instead, and more importantly for our purposes, following our earlier points about the general nature of scientific composition the Compositional reductionist will argue that pure macro-commonalities are nothing more than microphysical properties and relations, of aggregations of microphysical individuals, acting together – where, crucially, the reductionist will accept that heterogeneous combinations of microphysical properties and relations may *act together in the same ways.*

Given this stance, the ontological reductionist can adapt and endorse arguments about the indispensability of special science predicates offered by nonreductivists such as Kitcher. The point to mark is that both Kitcher and the Compositional reductionist each acknowledge the existence of aggregations of microphysical individuals with pure macro-commonalities. In this sense, they both accept what the sciences have shown us about the "organization of nature." And this shared commitment allows the Compositional reductionist to utilize Kitcher's argument for the indispensability of special science predicates. The ontological reductionist accepts there are macro-commonalities shared between heterogeneous aggregations and also agrees that lower-level predicates fail to capture these pure macro-commonalities. As a result, the Compositional reductionist concludes, like Kitcher and other recent nonreductivists, that special science predicates, referring to such pure macro-commonalities, are *in principle* indispensable in framing many explanations and other true statements. Of course, the key divergence of the Compositional reductionist from Kitcher, and other nonreductivists, is over the deeper nature of the pure macro-commonalities that they both take special science predicates to be necessary to capture. Using their battery of parsimony arguments, the reductionist argues that nonreductivists have erred by reifying commitments to such macro-commonalities into commitments to composed special science properties and individuals. For the Compositional reductionist claims that special science predicates are indispensable, in principle, only for referring to the ways of acting together shared by microphysically heterogeneous aggregations – where the Compositional reductionist argues this commits us only to microphysical properties/relations and individuals, albeit in a variety of heterogeneous aggregations.

We now have a clearer view of the commitments of Compositional reductionism and with it in hand we can finally return to the common concerns about this position, from both reductionist and nonreductivist philosophers of science, that we began this section by noting. If we start with the reductionist, then their concern focused upon whether an ontological reductionist could endorse theory nonreductivism. It is easy to see where such worries come from, for one driving intuition of reductionists is that all scientific statements, whether framed in the vocabulary of physics or using the different predicates of the special sciences, all have the same referents – the entities of physics. And reductionists have long assumed, swayed by the Nagelian account of reduction, that this point entails that there must be true identities statements showing that the predicates of physics and the special sciences have *exactly the same* referents. But we have now seen that even in a world that is structured as an ontological reductionist asserts, with just one layer of entities in those of microphysics, we can still fail to get true identity statements. When ontological reductionism is true, there need not be the tight connections amongst the predicates of higher and lower-level sciences needed for such identities, since the predicates of physics and those of the special sciences may refer to *different* features of the entities of physics and thus fail to neatly line-up in a way sufficient for true identity statements. This is just the situation that the Compositional reductionist envisions, for they argue that special science predicates refer to, and are indispensable for capturing, the pure macro-commonalities shared between heterogeneous microphysical aggregations. Thus although identity claims are unavailable, and despite the, in principle, indispensability of special science predicates, the Compositional reductionist still concludes that all the predicates of true scientific statements do only have the entities of physics as their sole referents, where predicates of various sciences refer to different features of such entities and their aggregations.

As we have noted already, that Compositional reductionism eschews identity statements is important because one of the main objections to "reductionism" is that identities are in fact usually unavailable. And we have now also found that Compositional reduction also avoids the other extant defenses of "nonreductivism" recently offered by philosophers of science. As we have seen in this section, neither arguments from multiple realization, nor those based upon the indispensability of special science predicates, undermine Compositional reductionism. One lesson of our findings is consequently that the famous recent defenses of "nonreductivism" have plausibly only made a compelling case for *theory* nonreductivism, but that this does not suffice to establish the truth of *ontological* nonreductivism. For we have seen that even a Compositional reductionist, who defends a comprehensive ontological reductionism, may endorse a robust form of theory nonreductivism based upon their acceptance of the phenomenon of multiple realization and the, in principle, indispensability of the predicates of the special sciences.[13] Thus, contrary to a common contention of nonreductivist philosophers, far from scientific practice trumping its "merely metaphysical" arguments, we have found that Compositional reductionism endorses the aspects of scientific practice rightly trumpeted by nonreductivists and then uses these features to drive its distinctive form of reduction. Compositional reduction is thus untouched by the extant, and rightly famous, defenses of "nonreductivism."

Our findings obviously have a range of wider implications for debates over reduction.[14] To take but one example, apparently as a result of the focus on Nagelian-style reduction, in recent discussions philosophers have often assumed that they have a dichotomy of options. *Either* be a "reductionist" who endorses some brand of theory reduction for the special sciences as well as a reduction in our ontology to the entities of microphysics. *Or* be a "nonreductivist" who accepts some brand of theory nonreductivism for special sciences as well as an unreduced ontology encompassing both the entities of physics and levels of composed special science entities. We have now found that the latter is a *false dichotomy*. For a Compositional reductionist accepts that our ontology reduces to the entities of microphysics, but nonetheless this reductionist may still endorse the, in principle, indispensability of special science predicates. This shows that the two strands of contemporary versions of "nonreductivism" and "reductionism" can come apart and were only seen as fused through use of the warped conceptual lens bequeathed to us by the Positivists' proscription of metaphysics in the philosophy of science. Compositional reduction thus raises a pressing question for many philosophers who label themselves either "nonreductivists" or "reductionists," since such writers must more carefully articulate the theses which are supported by their arguments, as well as the claims they actually intend to endorse.

To conclude, using our framework for the compositional relations posited in mechanistic explanations we have illuminated a novel form of ontological reductionism that has apparently long attracted working scientists. Unlike other forms of reduction, Compositional reductionism is impervious to objections based around the unavailability of property identities and the associated phenomenon of multiple realization which this form of reduction actually utilizes in its reductionist arguments. And Compositional reductionism is also untouched by recent defenses of the, in principle, indispensability of special science predicates which the Compositional reductionist again turns to their own purposes. Philosophers of science, and "naturalistic" philosophers more generally, thus need to speak to the challenge of Compositional reductionism, even though neither the legitimacy, or even the indispensability, of special sciences plausibly rides upon meeting this challenge.

Acknowledgments

Thanks to Jerry Fodor for very helpful discussion of the issues of this chapter. Special thanks also to Bob Richardson for pressing me over the years to think more carefully about the status of the concepts used in mechanistic explanations.

Notes

1 Figure 4.1 is drawn from Craver and Darden (2001).
2 For brevity I shall use "mechanistic" explanation to cover a large class of related explanations including, for example, "functional analyses" and "mechanism" or "mechanical" explanations. For a historical range of work on mechanistic explanation see, for example, Fodor (1968), Wimsatt (1974, 1976), Kitcher (1984), Bechtel & Richardson (1993), Machamer, Darden, & Craver (2000), and Craver & Darden (2001), amongst others.

3 See, for example, the papers in Kim (1993a) and especially more recent work such as Kim (1997, 1998, 1999).
4 By far the best existing introduction to the pioneering philosophical work in the general area of the metaphysics of science is provided by Brian Ellis (2002), though Ellis is not apparently aware of recent work on scientific composition by Kim and others.
5 See, for example, Gillett (2003a, 2003b, forthcoming a).
6 As will become clear below, there are a range of compositional relations between all manner of entities. As a result, I will use the term "composition" as the general term for such relations and I will provide specific names for particular compositional relations, between entities of certain categories, to avoid confusion amongst these relations.
7 The account is thus a relatively neutral version of Shoemaker (1980).
8 See, for example, Kitcher (1984, p. 25, n. 3) and Schaffner (1993, p. 287).
9 I defend a cruder version of the "thumbnail" account of realization in Gillett (2002, 2003a). A full account of comprising, realization, constitution and implementation, offered as parts of an integrated view of scientific composition generally, is presented in Gillett (unpublished a).
10 In fact, we get still further problems from the differing individuation and persistence conditions of higher and lower-level entities which again lead to the violation of Leibniz's Law. See Gillett (unpublished b) for a detailed discussion of the grave difficulties a deeper understanding of scientific composition raises for the existence of identities between higher- and lower-level scientific properties.
11 This argument underlies much work on the problem of mental causation and is what I have elsewhere called the "Argument from Realization" (Gillett & Rives, 2001; Gillett, 2003c, 2003b). Kim has perhaps done the most to illuminate the nature of this type of argument in a series of papers and books, see for example Kim (1997, 1998, 1999). Related arguments, including arguments directed at dispositional properties, are also found in Prior, Pargetter, & Jackson (1982), Martin (1997), and Heil (2000), amongst others.
12 It is important to note that the "structural," or "micro-based," properties feted by Kim (1998) and others are also consequently reduced by the New Reductionism. For if such structural properties exist at all, then our scientific evidence makes it clear that such structural are plausibly realized properties and hence the Extended Argument from Realization also applies to them. (See Gillett & Rives, 2001).
13 A few philosophers have defended such a position combining ontological reductionism and the indispensability of special science predicates. The clearest, and most detailed, examples are Schiffer (1987) and Loewer

(2001, unpublished), but a number of other philosophers may also be fruitfully interpreted in this fashion; see, e.g., Dennett (1991).

14 For the implications of Compositional reduction for debates over "emergence" see Gillett (forthcoming b).

References

Bechtel, W., & Richardson, R. C. (1992). *Discovering Complexity: Decomposition and Localization as Scientific Research Strategies.* Princeton: Princeton University Press.

Campbell, D. (1958). Common fate, similarity and other indices of the status of aggregates of persons as social entities. *Behavioral Science, 3,* 14–25.

Craver, C., & Darden, L. (2001). Discovering mechanisms in neurobiology: The case of spatial memory. In P. Machamer, R. Grush, & P. McLaughlin (Eds.), *Theory and Method in Neuroscience* (pp. 112–137). Pittsburgh, PA: University of Pittsburgh Press.

Dennett, D. (1991). Real patterns. *Journal of Philosophy, 88,* 27–51.

Ellis, B. (2002). *The Philosophy of Nature.* Montreal: McGill University Press.

Fodor, J. (1968). *Psychological Explanation.* New York: Random House.

Fodor, J. (1974). Special sciences, or the disunity of science as a working hypothesis. *Synthese, 28,* 97–115. Reprinted in N. Block (Ed.), *Readings in Philosophy of Psychology* (Vol. 1, pp. 120–133). Cambridge, MA: Harvard University Press, 1980, and in J. Fodor, *Representations: Philosophical Essays on the Foundations of Cognitive Science* (pp. 127–145). Cambridge, MA: MIT Press.

Fodor, J. (1975). *The Language of Thought.* New York: Crowell.

Gillett, C. (2002). The dimensions of realization: A critique of the standard view. *Analysis, 62,* 316–323.

Gillett, C. (2003a). The metaphysics of realization, multiple realizability and the special sciences. *Journal of Philosophy, 100,* 591–603.

Gillett, C. (2003b). Non-reductive realization and non-reductive identity: What physicalism does not entail. In S. Walter & H.-D. Heckmann (Eds.), *Physicalism and Mental Causation* (pp. 31–58). Exeter, UK: Imprint Academic.

Gillett, C. (2003c). Strong emergence as a defense of non-reductive physicalism: A physicalist metaphysics for "downward" determination. Special issue on emergence. *Principia, 6,* 83–114.

Gillett, C. (forthcoming a). Samuel Alexander's emergentism: Or, higher causation for physicalists. *Synthese.*

Gillett, C. (forthcoming b). The hidden battles over emergence. In P. Clayton (Ed.), *The Oxford Handbook of Religion and Science.* Oxford: Oxford University Press.

Gillett, C. (unpublished a). Making sense of levels in the sciences: Powers, properties, parts and processes.

Gillett, C. (unpublished b). Realization and identity in the sciences: Scientific composition and the unavailability of identities.

Gillett, C., & Rives, R. (2001). Does the argument from realization generalize? Responses to Kim. *Southern Journal of Philosophy*, *39*, 79–98.

Heil, J. (2000). Multiple realizability. *American Philosophical Quarterly*, *36*, 189–208.

Kim, J. (1993a). *Supervenience and Mind: Selected Philosophical Essays.* New York: Cambridge University Press.

Kim, J. (1997). The mind–body problem: Taking stock after forty years. *Philosophical Perspectives*, *11*, 185–207.

Kim, J. (1998). *Mind in a Physical World: An Essay on the Mind–Body Problem and Mental Causation.* Cambridge, MA: MIT Press.

Kim, J. (1999). Making sense of emergence. *Philosophical Studies*, *95*, 3–44.

Kitcher, P. (1984). 1953 and all that: A tale of two sciences. *Philosophical Review*, *93*, 335–373. Reprinted in P. Kitcher, *In Mendel's Mirror* (pp. 3–30). New York: Oxford University Press, 2003. (All references are to the reprint.)

Loewer, B. (2001). From physics to physicalism. In C. Gillett & B. Loewer (Eds.), *Physicalism and its Discontents* (pp. 37–56). Cambridge: Cambridge University Press.

Loewer, B. (unpublished). Why is there anything except physics?

Machamer, P., Darden, L., & Craver, C. (2000). Thinking about mechanisms. *Philosophy of Science*, *67*, 1–25.

Martin, C.B. (1997). On the need for properties: The road to Pythagoreanism and back. *Synthese*, *112*, 93–231.

Nagel, E. (1961). *The Structure of Science: Problems in the Logic of Scientific Explanation.* New York: Harcourt, Brace, and World.

Prior, E., Pargetter, R., & Jackson, F. (1982). Three theses about dispositions. *American Philosophical Quarterly*, *19*, 251–257.

Putnam, H. (1975a). Philosophy and our mental life. In H. Putnam, *Mind, Language and Reality: Philosophical Papers* (Vol. 2, pp. 291–303). Cambridge: Cambridge University Press.

Putnam, H. (1975b). The nature of mental states. In H. Putnam, *Mind, Language and Reality: Philosophical Papers* (Vol. 2, pp. 429–440). Cambridge: Cambridge University Press.

Schaffner, K. (1993). *Discovery and Explanation in Biology and Medicine.* Chicago: University of Chicago Press.

Schiffer, S. (1987). *Remnants of Meaning*. Cambridge, MA: MIT Press.
Shoemaker, S. (1980). Causality and properties. In P. Van Inwagen (Ed.), *Time and Cause* (pp. 228–254). Dordrecht: Reidel.
Simon, H. (1969). *The Sciences of the Artificial* (2nd ed.). Cambridge, MA: MIT Press.
Weinberg, S. (1992). *Dreams of a Final Theory: The Search for the Fundamental Laws of Nature*. New York: Pantheon.
Wimsatt, W. (1974). Reductive explanation: A functional account. In A. C. Michalos, C. A. Hooker, G. Pearce, & R. S. Cohen (Eds.), *PSA 1974* (pp. 671–710). Dordrecht: Reidel.
Wimsatt, W. (1976). Reductionism, levels of organization and the mind–body problem. In G. Globus, I. Savodnik, & G. Maxwell (Eds.), *Consciousness and the Brain* (pp. 199–267). New York: Plenum.
Wimsatt, W. (1994). The ontology of complex systems: Levels of organization, perspectives and causal thickets. *Canadian Journal of Philosophy, supp. vol. 20*, 207–274.

5

REDUCTIONISM, EMBODIMENT, AND THE GENERALITY OF PSYCHOLOGY

Lawrence A. Shapiro

1. Introduction

A central controversy in philosophy of psychology pits reductionists against, for lack of a better term, autonomists. The reductionist's burden is to show that psychology is, at best, merely a heuristic device for describing phenomena that are, when speaking more precisely, just physical. I say "at best," because reductionists are prone to less conciliatory remarks, such as: "psychological property P just is physical property N, so scientific explanation might as well focus exclusively on N," and "psychological property P is nothing other than N, so generalizations about N suffice to say all that there is to say about P," and "knowledge of all the N facts suffices for knowledge of all the P facts."

The autonomist's burden, on the other hand, is to explain why psychology should not be reduced to physics, given the naturalistic assumption that any two objects alike in all their physical properties will also be alike in all their psychological properties, and that these physical properties completely determine the psychological properties.[1] If the naturalistic assumption is true, and most autonomists *do* accept it,[2] then the task of defending a coherent conception of psychology's autonomy becomes hard indeed. As matters stand, it certainly looks as if a complete physical reckoning will say all that needs to be said about psychological phenomena.

Traditionally, autonomists have met their burden by appeal to the purported multiple realization of psychological states.[3,4] Because, they claim, psychological properties can be realized by different kinds of physical properties, and disjunctions of physical properties are not themselves physical properties, psychology cannot be reduced to physics. It is false that psychological property P is nothing other than physical property N. Generalizations about N do not suffice to say all that there is to say about P. Knowledge about N facts does not suffice for knowledge about P facts. Psychology is autonomous because psychological kinds are invisible from the perspective of physics. Whatever psychological generalizations are available will emerge from the study of these kinds, and physics has no place in this investigation.

Familiarity with the reduction literature leaves the impression that autonomists see themselves as the good guys. The multiple realization argument is a Declaration of Independence and the autonomists are fighting to maintain the sovereignty of the special sciences. However, this posturing is somewhat mystifying. Reductionist pursuits have not been undertaken with the goal of shutting down valuable scientific programs. Rather, the point of reduction is to gain generality and unification (Nagel, 1961). Reduction buys generality, for if it is true, say, that water *is* H_2O, then whatever is true of one sample of water will be true of all samples of water. Similarly, if water is H_2O we then have an explanation for why water exhibits a variety of properties whose association might otherwise appear accidental (freezing point, boiling point, density, specific gravity, and so on). Surely the generality and unification that reduction provides is valuable, and thus autonomy, if it truly does entail a loss of these things, is not so clearly a desirable end.

Jaegwon Kim (1992, 1998b) especially has emphasized the cost of autonomy. According to Kim, if higher-level properties are multiply realizable, then generalizations at the higher level are possible only within narrow confines. For instance, he denies that there can be psychological generalizations that cross species lines. Human psychology and Martian psychology, assuming that human beings and Martians realize their psychological processes in distinct physical ways, could share no laws. This follows, Kim thinks, because differences in psychological realization prevent similarities at the psychological level. Differences, as it were, *trickle up*. Kim's conclusions are important because they directly challenge the presumed benefit of multiple realization. Multiple realization, he argues, may bring autonomy, but the autonomy brings with it an acute constriction in domain. Given my previous remarks about the goals of reduction, perhaps it would not be so surprising if the troubles that multiple realization makes for reduction also cause strain in the idea of an autonomous *and* general psychology.

In this chapter I consider two challenges to the generality of psychology. The first – Kim's – depends on an argument that I do not think is sound. However, I want also to consider a body of psychological research that suggests an independent reason for thinking that psychology cannot be general. This research, sometimes labeled *embodied cognition*, may tell against the possibility of psychological generalizations that span species. If certain claims that embodied cognition theorists make are correct, I think the prospects for a general psychology are slim, but for reasons other than those that emerge from the multiple realization thesis. Kim's vision of a fractured psychology may be accurate, but not for his reasons.

2. Multiple Realization and Autonomy

Autonomy, in this context, is the idea that the special sciences have explanations, kinds, and subject matters that are unavailable or inaccessible to physics. For instance, consider the psychological predicate "believes that the lake is frozen." Defenders of autonomy hold that this predicate is not simply another name for a physical property. The psychological property is not identical to any physical property, and so explanations that appeal to the psychological property have no physical counterpart. Thus, even a completed physics could not replace psychology. Physics is about the physical domain; psychology is about the psychological domain; and never the twain shall meet.

As I mentioned above, the justification for belief in autonomy comes from the apparent multiple realization of psychological kinds. The human belief that the lake is frozen is realized in a brain state, whereas the Martian's and robot's beliefs that the lake is frozen are realized in different kinds of physical states. The motivation for claiming that the human being, Martian, and robot all have the same kind of belief is the functional similarity these agents exhibit. Each agent, for instance, if confronted with the same stimuli, forms the same desires, engages in the same actions. Counterfactuals true of one, e.g., if she had not believed that the lake was frozen then she would not have stepped onto it, are true of the others. The three are psychologically identical despite their physical differences.

To summarize, the connection between multiple realization and autonomy rests on the following argument.

1. Psychological properties are multiply realizable.
2. If psychological properties are multiply realizable then psychological generalizations over these properties have no physical counterparts.
3. Psychological generalizations have no physical counterparts.
   Therefore, psychology is autonomous.

3. Disintegration from Below

Jaegwon Kim (1992, 1998b) has argued that the multiple realizability on which the defense of autonomy rests is, in fact, a cause for the disintegration of psychology as well as the other special sciences.[5] His argument begins with two principles:

> *Causal Individuation of Kinds*: Kinds in science are individuated on the basis of causal powers, i.e., objects share a property insofar as they have similar causal powers. (1992, p. 17)
>
> *Causal Inheritance*: If mental property M is realized in a system at *t* in virtue of physical realization base P, the causal powers of this instance of M are identical with the causal powers of P. (1992, p. 18)

The first principle, Kim thinks, is "plausible," and, "in any case, widely accepted" (1992, p. 17). Of course, this is not to say that all philosophers would accept it, and those who have defended externalist theses about mental content, e.g., Burge (1989), may well reject it. Nevertheless, the principle does have intuitive appeal. If instances of two different kinds share all their causal powers, one must wonder why it is necessary to distinguish the kinds. Certainly the kinds could not be experimentally distinguished, but then by what other criteria might a scientist draw the distinction?

The second principle also aligns neatly with intuition. If a vase is realized in glass, then it will break just when the glass will break, will melt just when the glass will melt, will weigh just what the glass weighs. The vase, in other words, will have all and only the causal powers of the glass that realizes it. To deny this, Kim thinks, commits one to the existence of "causal powers that magically emerge at a higher level and of which there is no accounting in terms of lower-level properties and their causal powers and nomic connections" (1992, p. 18).

I propose to grant Kim his principles. More interesting than the principles is their implication. Kim believes that acceptance of the principles leads to the fracturing of psychology into a collection of individual, species-local, psychological theories. I shall call Kim's argument for this conclusion *Fracture*. The first premise of Fracture is this:

(1) Mental kind M is multiply realized by distinct physical kinds P_h and P_m

This premise is tantamount to a statement of autonomy. Kind M is a kind despite its multiple realization because, presumably, there are generalizations involving kind M that cannot be captured in the language of the realizers' domains. From the perspective of the science of the realizers, these higher-level generalizations are invisible. This is why psychology is indispensable.

Premises (2)–(4) spell out some consequences of Kim's principles:

(2) By Causal Individuation of Kinds, the causal powers of P_h must differ from the causal powers of P_m
(3) By Causal Inheritance, the mental state P_h realizes cannot have all the same causal powers as the mental state P_m realizes.
(4) Thus, again by Causal Individuation of Kinds, P_h and P_m cannot realize the same mental state.

Premise (4) captures the "trickle-up" effect. Differences in the causal powers of the realizers trickle up to create differences in the realized kinds. Given the trickle-up effect, it follows:

(5) Therefore, "[e]ach mental kind is sundered into as many kinds as there are physical realization bases for it, and psychology as a science with disciplinary unity turns out to be an impossible project." (1992, p. 18)

(6) Hence, M cannot be multiply realized (from the contradiction between 1 and 5).

As stated, Fracture is a *reductio*. From the assertion of multiple realization, which, recall, is a statement of autonomy, and Kim's two principles, it follows that multiple realization of psychological kinds is in fact impossible. Far from a justification for autonomy, multiple realization proves to be the undoing of the special sciences.

As I stated earlier, I do not think that Fracture is sound. This is because premise (4) – the trickle-up premise – is false. Underlying this premise is an unidentified principle, which I shall call *Relevance*:

> *Relevance*: all causal powers of a realizer are relevant to the individuation of the kind that it realizes.

Relevance justifies the trickle-up effect. If Relevance is true, then because P_h and P_m differ in their causal powers, the kinds that they realize must differ as well. Clearly, however, Relevance is not true. Consider the corkscrews I discussed (Shapiro 2000). The steel and aluminum double-lever corkscrews differ in many of their causal powers in virtue of their difference in realization. But, relative to their end as corkscrews, the fact that one might bend under 2,000 lb of pressure and the other under 3,000 lb, or that one might melt at a higher temperature than another, and so on, is irrelevant. Indeed, one might plausibly hold that no two instances of a functional kind are ever realized identically and therefore adoption of Relevance renders the notion of a functional kind incoherent. For instance, two steel corkscrews, one with a molecular impurity that the other lacks, must, according to Relevance, be different kinds.

But one doesn't have to make up examples to see the problem with Relevance. It is routine in science that schemes of individuation cross-classify each other. Such cross-classification is possible because the causal powers of interest in some scientific domains can be ignored in others. The chemist might be interested in DNA for its acidic properties. The geneticist's focus on DNA traces to its suitability as a unit of heredity. Perhaps the chemist doesn't even know that DNA plays this biological role; and the geneticist is ignorant that DNA is electron-deficient. The causal powers that make DNA an acid are of no concern to the geneticist and the causal powers that make DNA the unit of heredity do not matter to the chemist. The point is that not all causal powers of a kind are relevant to the kind's individuation. DNA can simultaneously instantiate different kinds – acid, unit of heredity – because it has causal powers that are both relevant and irrelevant to its various individuations.

Moreover, it is precisely because Relevance is false that chemists can say of both DNA and HCl that they are acids. The discovery that DNA plays a role in heredity did not force the chemist to jettison it from his list of acids. Acids can and do differ in their causal powers. This is OK and consistent with what we know to be true: objects can instantiate many different kinds all at once.

So, what happens to Fracture once we recognize that the same object may instantiate a property in the domain of one science but not another (e.g., HCl counts as a kind in chemistry but not in genetics) and that the same object may count as one kind in one science but a different kind in another (DNA is an acid in chemistry but not a gene, and it is a coder for proteins in genetics but not an acid)? If Relevance is false then the trickle-up premise loses its force. There is no longer a reason to think that differences in realization must trickle up to differences at higher levels, and so no reason to suspect that the multiple realization of M is impossible.

Without (4), Fracture's conclusion no longer follows. To deny trickle-up is to embrace the possibility that, relative to psychology, a set of distinct physical kinds $\{R_1, R_2, \ldots, R_n\}$ can realize a single mental kind. Given this, psychology can be at once autonomous and also general. Psychological generalizations are free to range over different species because physical differences in species do not entail psychological differences.

4. Disintegration from Above

Kim locates the fracturing of psychology in the alleged fact of multiple realization. I have just argued that multiple realization does not lead to this conclusion. However, there may be another reason to question whether psychology can be general or, alternatively, whether psychology must face disintegration. In the remainder of this chapter I would like to consider a motivation for psychological disintegration that comes from the field of embodied cognition (henceforth EC). This research suggests its own kind of challenge to a general psychology. I will call the argument that presents this challenge *Body Determinism*:

(1) EC lends support to the idea that the body and environment are constitutive of cognition.
(2) If EC lends support to the idea that body and environment are constitutive of cognition, psychology must partition its domain into species-specific subject matters. Therefore, psychology must partition its domain into species-specific subject matters.

Body Determinism differs considerably from Fracture. Fracture depends on a couple of metaphysical principles about individuation and causal inheritance, as well as a tacit premise about the role of relevance in individuation. Body Determinism shares Fracture's conclusion, but, unlike Fracture, it is ultimately empirical. To be sure, philosophical considerations will be prominent in the argument. In particular, questions about individuation remain important. However, unlike in the case of Fracture, the motivation for Body Determinism derives from a corpus (no pun intended) of empirical work.

Also distinguishing the arguments is their intended breadth. Kim thinks Fracture applies not only to psychology, but to any science that includes functional kinds in its domain. On the other hand, Body Determinism is just about psychology. This difference in scope reflects the difference that empirical considerations make in each argument. Fracture has an *a priori* feel, and confirming this is the fact that the empirical details of a science make no difference to whether Fracture applies to it. Any science, if it has multiply realizable kinds, will end up fractured. However, because Body Determinism takes its motivation from data about how the mind works, it has no application to sciences that are not about minds.

With preliminary remarks behind us, it is time to consider the first premise of Body Determinism. The following quotations from philosophers and psychologists, while not exactly clarifying the first premise, at least give it expression in ways that can be further illuminated.

> [T]he peculiar nature of our bodies shapes our very possibilities for conceptualization and categorization. (Lakoff & Johnson, 1999, p. 19)

> If perception is in part constituted by our possession and exercise of bodily skills – as I argue in this book – then it may also depend on our

> possession of the sort of bodies that can encompass those skills, for only a creature with such a body could have those skills. (Noë, 2004, p. 25)

> The cognitive processing that gives rise to mental experience may be something whose functioning cuts across the superficial physical boundaries between brain, body, and environment. (Spivey, Richardson, & Fitneva, 2004, p. 178)

> I will argue that many cognitive capacities in symbol using creatures, far from being purely internal, are either *enactive bodily* capacities, or *world-involving* capacities. These capacities are not realized by some internal arrangement of the brain or central nervous system, but by embodied states of the whole person, or by the wide system that includes (parts of) the brain as a proper part. (Wilson, 2004, p. 188, original emphasis)

Common to these passages is the idea that an organism's body shapes an organism's mind or mental capacities. This idea of shaping, however, requires scrutiny. There are many ways that an organism's body might shape its mind that have no special interest. An empty stomach might cause an organism to desire food. Having two eyes allows depth perception by the detection of disparities in the retinal images. Having a head that can rotate on a neck opens perceptual opportunities that a stationary head would lack.

These examples of shaping, however, all point to the body as a cause of or influence on cognitive capacities. But, of course, it has long been standard in cognitive science to grant the body this kind of role in cognition. A cognitive scientist wedded to the idea that the mind is realized entirely in the brain, or supervenes entirely on the brain, will explain the above examples by appeal to a notion like transduction. The empty stomach causes a desire for food only in virtue of a process by which the stomach's emptiness becomes represented in a code that the brain can understand. Similarly, the retinal images or the motions of the head can cause or influence perception only by their translation into a neural code. Once head motion, for instance, becomes encoded, it is accessible to the cognitive processes in the brain that result in perception. According to this internalist picture, the body does shape cognition, but only by a process of transduction that converts bodily goings on into a form on which the brain can operate.

Of course, the authors above recognize all this.[6] They see the body as something much more than just a causal influence on the mind. Lakoff and Johnson, for instance, argue that human beings are equipped with a class of basic concepts that, through metaphorical reasoning, can give rise to more sophisticated concepts. The concept of HAPPY, they speculate, is rooted in the basic concept UP; SAD derives from DOWN. Through application of metaphorical reasoning, UP and DOWN structure the concepts HAPPY and SAD: "I'm feeling up. That boosted my spirits. My spirits rose. You're in high spirits. Thinking about her always gives me a lift. I'm feeling down. I'm depressed. He's really low these days. If fell into a depression. My spirits sank" (1980, p. 15). Crucially, the reason HAPPY and SAD are structured in terms of UP and DOWN has to do with the nature of our body. "Drooping posture," Lakoff and Johnson suggest, "typically goes along with sadness and depression, erect posture with a positive emotional state" (1980, p. 15).

Because my concern is less with the tenability of Lakoff and Johnson's claims than it is with their consequences for the nature of psychology, I do not intend to discuss them critically here. For present purposes, Lakoff and Johnson's work suggests a sense in which the body shapes cognition that goes beyond the banal claim that the body has a causal influence on cognition. If Lakoff and Johnson's account of concept acquisition is correct, it follows that the body informs cognition by constraining how an agent can conceive the world. This is perhaps easiest to see in a fanciful case: "[i]magine a spherical being living outside of any gravitational field, with no knowledge or imagination of any other kind of experience. What could UP possibly mean to such a being?" (1980, p. 57). With no concept of UP available to this being, its concepts of HAPPY and SAD would be nothing like our own. If possible at all, they would be structured by entirely different metaphors.

One might insist that the body's properties are indeed causes of the concepts HAPPY and SAD, but I hope it is clear that, if so, they are not causes in the same sense that an empty stomach or a moving head contributes causally to cognition. In the latter cases, the empty stomach and the moving head are encoded in a fashion that allows the information they carry to be integrated into a computation resulting in a desire for food or a perception of (perhaps) depth. But in the former case the body is a cause in the way that a spout's shape is a cause of the volume of liquid it can pour in a given amount of time. The spout's shape constrains the flow of liquid but it needn't be encoded to have this effect on the volume of flow. Just as spouts that differ in shape will differ also in the volume of liquid they allow to pass in a given time, so, according to Lakoff and Johnson, organisms that differ in bodily shape will differ as well in the kinds of concepts they can acquire.

The spout analogy is helpful in another sense. Clearly two spouts that differ in the volume of liquid they pour must differ as well in other properties. That is, their pouring capacity supervenes on properties like their size, surface texture, and so on. We might take Lakoff and Johnson's discussion of the spherical being to imply that concept acquisition supervenes not just on neural properties, but also on bodily properties. This makes clear the difference between the body's being a causal influence on the mind and the body's being something more like a constituent of the mind. Concepts like HAPPY and SAD, if Lakoff and Johnson have correctly described the manner of their acquisition, supervene not just on the brain but on the body as well. Of course, this does not imply that differences in body must entail differences in concepts, but it does reveal that differences in body are the kinds of differences that can suffice for differences in concepts.

Noë (2004) also makes a case for conceiving of the body as a constituent of cognition. He defends a skill-based account of perception, according to which perception is possible only in virtue of acquaintance with a special class of sensorimotor skills. The difference between a skill-based account of perception and traditional representational account can be brought out with the following example. Consider two strategies for moving through narrow passageways without banging your head on the walls to either side. On a traditional account, one forms a representation of the passageway, a representation of one's position relative to the passageway, and computes a course that will take one safely through. In effect, once the representations have been constructed, one could proceed with one's eyes closed. On the skill-based account of perception that Noë favors, one might orient oneself in front of the passage so that the amount of opening to the left and right always appears to be the same. This insures that one is moving toward the center of the opening. This second strategy does away with or minimizes the need for computations over representations, demanding instead the exercise of a particular skill.

On the skill-based model, one forsakes representation-guided computation for world-guided activity. Aligning oneself with the center of the passageway requires constant calibration with the amount of open space that appears to one's left and right. If more space appears to the right, one must move toward the right in order to create more open space on the left. This kind of interaction with the world constitutes a kind of knowledge. One knows, tacitly at least, that one will pass through the center of the doorway if one evenly balances the amount of opening that appears to the left and right. This knowledge, however, leads to success for purely contingent reasons. For one thing, it would not work for organisms whose bodies were extensively asymmetrical. An organism whose head was fastened to the right side of its body would deploy the strategy only at risk of bruising its left side. Likewise, if the world were such that passageways were vast, the strategy would seldom be of practical value.

I have made up the above example to illustrate the flavor of the theory of perception that Noë wishes to defend (but, for all I know, human beings really do adopt such a skill-based strategy to navigate successfully through doorways). One can take Noë's project as an effort to see how far skill-based accounts can go in explaining all of perception and cognition. Insofar as perception and cognition are skill-based, Noë believes that this makes the body intrinsic to psychology:

> In general it is a mistake to think that we can sharply distinguish visual processing at the highly abstract algorithmic level, on the one hand, from processing at the concrete implementational level, on the other. The point is not that algorithms are constrained by their implementation, although that is true. The point, rather, is that the algorithms are actually, at least in part, formulated *in terms of* items at the implementational level. You might actually need to mention hands and eyes in the algorithms! (2004, p. 25, his emphasis and excitement)[7]

It is tempting to think that Noë overstates his case here. Why is it necessary to mention hands and eyes in the algorithms involved in visual processing? I can imagine two ways to resist this claim. First, one might insist that hands and eyes can affect visual processing only by virtue of an encoding of the information they provide. It is something like a category mistake to think that an eye – a spherical chunk of proteins – can be part of an algorithm that takes as input light reflected from surfaces and produces as output a visual scene of the world. How can an eye "get into" such an algorithm? On the other hand, there is no problem imagining that representations of the eye's properties can participate in an algorithm. If, say, the eye's motion contributes something to perception, this can be only because this motion is represented in a code that plays a role in the visual processing algorithm.

Second, one might see hands and eyes as merely *facilitating* visual perception. Movements of the eyes and hands might facilitate the perception of occluded objects, or distance, or shape. But this is not to say that eyes and hands are literally part of visual perception. They are no more part of visual perception than eyeglasses are. They provide opportunities that make possible, or in some way enrich, the processes that yield visual perception (Clark, forthcoming). On this view, Noë errs in extending the realm of psychology beyond its proper borders. He mistakes the track for part of the train just because trains can't move without tracks.

Perhaps neither of these objections to Noë are easy to dismiss. However, it is clear that there is a fair amount of question-begging going on. For instance, to deny that eyes and hands are part of the visual process already assumes something about the nature of visual processing. Noë's plaint is that vision, as well as other psychological capacities, has been misconceived. Hands and eyes might be part of vision if one takes vision to be a skill-based activity – an activity that requires the use of sensorimotor knowledge. This response seems also to meet the challenge of the first objection. To insist that visual processing can make use only of suitably encoded information is again to prejudge the nature of psychological processing.

As with all disputes that involve charges of question-begging, the winner will be the one who makes his position superior on independent grounds. If Noë's skill-based account of cognition answers questions that representationalist accounts do not, if it resolves puzzles, makes predictions that would be surprising on more traditional accounts, and so on, then this is reason to take the account very seriously. Of course, I am not in a position to decide this issue. Fortunately, as was the case in my discussion of Lakoff and Johnson's work, I am less interested in whether skill based accounts will hold up in the long run than I am with tracing their consequences for the possibility of a unified psychology. I will turn to this issue after some further remarks about how to understand premise (1) of Body Determinism.

The quotations from Spivey et al. (2004) and Wilson (2004) suggest another sense in which the body or environment is constitutive of cognition. I think Wilson is clearest here. Wilson describes processes of *wide computation*. Computation is wide when it involves representational or informational states that are external to the brain. Examples are easy to come by. Hutchins (1995) describes the navigator's use of navigational instruments like a sextant to compute location. Wilson discusses the use of pencil and paper to aid in arithmetic tasks. Clark and Chalmers (1998) ask us to consider the role of a pocket diary in an Alzheimer's victim's interaction with the world. Each of these cases involves a process in which an external prop plays a computational role.

Consider, for instance, the use of pencil and paper in computing the product of 683 and 548. One could do this computation without benefit of external aids. However, most of us would have an easier time with the task if we could "offload" steps in the computation to a piece of paper. The marks on paper are, in a completely literal sense, parts of the multiplication process. The steps involved in this process are realized in not just the brain, but in the brain and the marked paper.

Clark and Chalmers (1998) provide a similar example. Otto, who has Alzheimer's disease, relies on his notebook to provide him with the information that MoMA is on 53rd St. This information in his notebook, Clark and Chalmers claim, can play the same causal role for Otto that Inga's in-the-head belief that MoMA is on 53rd St. plays for her. The moral, they say, "is that when it comes to belief, there is nothing sacred about skull and skin. What makes some information count as a belief is the role it plays, and there is no reason why the relevant role can be played only from inside the body" (1998, p. 14).

The point to which proponents of embodied cognition wish to draw attention is that once one gives up a reductionist view of mind – a view of mind that identifies mental properties with physical properties – and adopts instead a functional conception of mental states, one cannot avoid the consequence that things external to the brain might qualify as mental states. If the criteria for mental states are specified functionally, then there is nothing to prevent bodily or environmental states from being mental. There is, in other words, no reason to align the mental boundaries with the neural boundaries. The supervenience base of the mind, or, at any rate, cognitive processes, can include items as diverse as notebooks, fingers, calculators, PDAs, and memory sticks.

Perhaps at this point I have said enough to convey the gist of premise (1) of Body Determinism. Again, my goal is not to defend premise

(1), but to consider its consequences for the science of psychology. With this in mind, it is time to consider premise (2) – the claim that embodiment implies that psychology must fracture itself into as many sciences as there are relevantly different species.

Suppose it is true that cognition is embodied in the way that Lakoff and Johnson (1980) surmise, i.e., that anatomical facts constrain and influence the contents of thought, and that the contents of thought supervene on these anatomical facts. Also suppose that perception is best conceived as intrinsically sensorimotor, requiring skillful interactions between body and environment, as Noë believes. Further suppose it is true that much of cognition is wide in the sense that Spivey et al. (2004), Wilson (2004), and Clark and Chalmers (1998) describe. Beliefs, on such a view, can exist outside the head and, indeed, outside the body. The mind's realization can extend beyond the brain and into the world. How do these points bear on the nature of psychology?

One might take the claims from EC as supporting a view of psychology very much like the one that Fracture envisions. Just as Kim thinks that psychology must disintegrate into species-specific sciences because differences in realization trickle up to create differences in psychology, a proponent of EC might think that because the kind of body an organism has makes a difference to the kind of psychology it has, psychology must splinter itself into as many subdisciplines as there are types of body.

At this point the difference between the sort of claim that Noë and Lakoff and Johnson make and that which Wilson and Clark and Chalmers make grows in prominence. If differences in body type are to suffice for the fracturing of psychology, then the body must have a role in cognition that is deeper than mere realization. For, consider again Clark and Chalmers's claim that the entries in Otto's notebook qualify as beliefs. The justification for this claim is that the notebook entries are functionally identical to beliefs. Hence, if one is a functionalist, then this stipulation makes it trivial that the notebook entries are beliefs. However, functionalism supports the view that psychology can be general. The functionalist is untroubled by the failure of reduction and takes psychological generalizations to range over mental states that are individuated in terms of their causal role in a network of inputs, other mental states, and outputs. Wide computationalists like Wilson and Clark and Chalmers are simply functionalists who have come to appreciate that the states which play "mental" causal roles need not all be in the head.[8]

On the other hand, Lakoff and Johnson and Noë suggest that bodily properties are sometimes constitutive of cognition. To have concepts like HAPPY and SAD that bear any resemblance to these concepts as human beings understand them *requires* a human-like body. The human body is not just the realization base for these and other concepts, but it is in virtue of having a human body that human concepts mean what they do.[9] Or, for Noë , perception is the disposition to respond to features of the environment in ways particular to a certain sort of body. Having a different sort of body necessitates a different kind of perception, hence, "only a creature with a body like ours can have experiences like ours" (Noë, 2004, p. 26).

An analogy may be useful for clarifying Lakoff and Johnson and Noë's view. Consider Whorf's principle of linguistic determinism. Stated roughly, this principle claims that the structure of the language one speaks determines the kinds of thoughts one can have. Individuals who speak languages with distinct grammatical categories may as a result conceptualize the world differently. My interest is not in the truth of Whorf's principle, but in its implication for psychology. If the principle is true, a psychologist interested in how an individual conceives his world would have to attend to the structure of an individual's language. Insofar as an individual's mind bears the stamp of his language, psychology cannot proceed independently of linguistics.

One can usefully think of Body Determinism as linguistic determinism writ large. Insofar as an individual's mind bears the stamp of its body, psychology cannot proceed independently of an investigation of the body. But Body Determinism has more significant consequences for psychology than does linguistic determinism. Linguistic determinism does not entail the fracturing of psychology because it does not challenge the idea that members of differing linguistic communities may nevertheless share some psychological capacities. Indeed, this must be so, for the thesis of linguistic determinism is interesting only if differences in some psychological capacities, e.g., categorization, are the product of linguistic differences rather than the result of more fundamental psychological differences. Hence, linguistic determinism presumes a level of psychological uniformity.

However, Body Determinism is not so committed. If Lakoff and Johnson and Noë tell a coherent story of the body's influence on the mind, then there need not be any universal psychological capacities. In such an event, psychology has no choice but to give up hope of generality in favor of species-specific regularities that bear all the idiosyncrasies of the organisms over which they range. Psychology disintegrates into myriad subfields, each dedicated to the psychology of a particular species.

5. Conclusion

The relationship between reduction and autonomy is complicated. Functionalists see autonomy as emerging from the failure of reduction. Reduction fails because psychological kinds are multiply realizable; and because reduction fails psychology is autonomous. However, because reduction carries with it the promise of unification and generality, one might worry that, as it goes, so go these virtues. Kim's Fracture argument develops this worry. However, I have argued, Fracture fails because it rests on a relevance assumption that is false. Kinds exhibit a variety of causal powers, and the individuation of kinds needn't take into account all of them. This means that differences between realizers don't necessarily trickle up into differences at higher levels. Different chemical kinds can realize the same psychological kinds. Consequently, reduction can be false while a general psychology remains possible.

But what if, as some researchers in EC have argued, psychological processes are embodied in the sense that they intrinsically comprise bodily processes? I have argued that this might lead to the disintegration of psychology. No longer could psychology generalize over differently embodied organisms if these differences are constitutive of differences in psychology. For each species there would be *sui generis* psychological laws. This consequence, I think, sets a course into uncharted territory. Philosophers have tended to see the choices between reduction and autonomy as mutually exclusive but also as exhaustive. Reductionism is either true or false, and psychology is autonomous if only if it is false. But the kind of disintegration of psychology that EC suggests embraces neither of these options. Here we have reduction *of a sort*. Psychological identity implies bodily identity. But we also have autonomy *of a sort*. Organisms that differ physically may nevertheless be psychologically similar if they share the relevant morphological and physiological properties.[10]

I wish to emphasize again that I have not attempted to defend the conception of psychology that leads to what I have called Body Determinism. I have merely sought to reveal consequences that EC may have for the unity of psychology. So, are the claims of EC that lead to the disintegration of psychology true? Without access to alien beings, the claims are difficult to test. Perhaps work on nonhuman animals may someday bear results. On the other hand, the principle of linguistic determinism remains controversial to this day, despite ready access to members of distinct linguistic communities.

Notes

1 Kim notes frequently that a minimal sort of physicalism requires at least both these claims, as supervenience by itself does not imply dependence (see, for instance, 1998a, pp. 10–12).

2 With some codicils about wide content.

3 See Shapiro (2004) for some reasons to be cautious about the multiple realization assumption.

4 There is another kind of burden that autonomists face: the risk of causal exclusion. I intend to put this to the side in this chapter, but for discussion see Kim (1998b) and Shapiro and Sober (forthcoming).

5 Discussion in this section draws heavily on Shapiro (forthcoming).

6 But perhaps they are not always as careful as they should be. Block (2005) takes Noë to task for sometimes confusing the claim that the body is a causal influence on the mind with the claim that the body is a constituent of the mind.

7 I make a similar point in Shapiro (2004), where I draw an analogy between, on the one hand, the doomed effort to pilot an airplane through use of instructions designed for a submarine and, on the other hand, the use of human psychological algorithms to guide action in a nonhuman body. The analogy was intended to show the sense in which human psychology is tailored to the particulars of a human body.

8 I am grateful to Andy Clark for helping me to see this. For valuable discussion of these issues, see Clark (forthcoming).

9 Clark (forthcoming) challenges this very strong claim about the body's role in cognition, arguing that the body is just one facet in the constituents of cognition. Perhaps, he suggests, compensatory changes in environment and internal processing can endow differently embodied organisms with a similar cognitive profile. This seems to me a fruitful line of criticism against Lakoff and Johnson.

10 Here I assume that differences between species entail differences between body types sufficient for differences in psychology. Perhaps species with very similar bodies will have very similar psychologies.

References

Block, N. (2005). Review of Alva Noë, *Action in Perception*. *Journal of Philosophy*, *102*, 259–272.

Burge, T. (1989). Individualism and causation in psychology. *Pacific Philosophical Quarterly*, *70*, 303–322.

Clark, A. (forthcoming). Pressing the flesh: Exploring a tension in the study of the embodied, embedded mind. *Philosophy and Phenomenological Research.*

Clark, A., & Chalmers, D. (1998). The extended mind. *Analysis*, *58*, 7–19.

Hutchins, E. (1995). *Cognition in the Wild*. Cambridge, MA: MIT Press.

Kim, J. (1992). Multiple realization and the metaphysics of reduction. *Philosophy and Phenomenological Research*, *52*, 1–26.

Kim, J. (1998a). *Philosophy of Mind*. Boulder: Westview Press.

Kim, J. (1998). *Mind in a Physical World: An Essay on the Mind–Body Problem and Mental Causation*. Cambridge, MA: MIT Press.

Lakoff, G., & Johnson, M. (1980). *Metaphors We Live By*. Chicago: University of Chicago Press.

Lakoff, G., & Johnson, M. (1999). *Philosophy in the Flesh: The Embodied Mind and Its Challenge to Western Thought*. New York: Basic Books.

Nagel, E. (1961). *The Structure of Science: Problems in the Logic of Scientific Explanation*. New York: Harcourt, Brace, and World.

Noë, A. (2004). *Action in Perception*. Cambridge: MIT Press.

Shapiro, L. (2000). Multiple realizations. *Journal of Philosophy*, *97*, 635–664

Shapiro, L. (2004). *The Mind Incarnate*. Cambridge: MIT Press.

Shapiro, L. (forthcoming). Can psychology be a unified science? *Philosophy of Science.*

Shapiro, L., & Sober, E. (forthcoming). Epiphenomenalism: The do's and the don'ts. In G. Wolters & P. Machamer (Eds.), *Studies in Causality: Historical and Contemporary*. Pittsburgh: University of Pittsburgh Press.

Spivey, M., Richardson, D., & Fitneva, S. (2004). Thinking outside the brain: Spatial indices to linguistic and visual information. In J. Henderson & F. Ferreira (Eds.), *The Interface of Vision, Language, and Action* (pp. 161–189). New York: Psychology Press.

Wilson, R. (2004). *Boundaries of the Mind: The Individual in the Fragile Sciences*. New York: Cambridge University Press.

PART II

PHILOSOPHICAL ACCOUNTS OF REDUCTIONISM, MECHANISM, AND CO-EVOLUTION

6

REDUCTION WITHOUT THE STRUCTURES

Robert C. Richardson

1. The Vision of the Two *Structures*

Ernest Nagel's *The Structure of Science* (1961) and Thomas Kuhn's *The Structure of Scientific Revolutions* (1962) are certainly two of the greatest works in twentieth-century philosophy of science. Published but a year apart, these books are representative of two very different trends in the philosophy of science. Discussions of each have emphasized their differences. They certainly differ in how they see the development of science. They differ in how they understand the relation of theories. Nagel sees science as conservative. Reduction becomes the index of success, since it is a measure of consolidation and increasing empirical success. Progress is understood as a matter of slow accumulation, and unification. Kuhn sees science as revolutionary. Revolution becomes the hallmark of radical progress. New visions are better visions even if they overturn the old world order. Despite their differences, there is much these two *Structures* share. Both arose within a Positivist tradition. Both offered a vision of science. Both recognized how important reduction is to science. Both understood science in terms of the structure of theories. And both were grounded in a common view of the structure of theories. Theories were seen as axiomatically organized structures, and explanations were deductions within these structures.

I think that what is shared by the two *Structures* is at least as important as where they differ. It is against the shared views that their differences stand out in relief. The differences between them suggest, among other things, a common flaw in how they conceive of reduction.

I also think that the difficulties they face is largely inherited by more recent work on reduction and reductionism. Despite the considerable sophistication and modification we have seen in the treatment of reduction since the *Structures*, the core vision is largely retained.

Nature is hierarchically organized. Herbert Simon captures that idea under the rubric of the "architecture of complexity" (1969, ch. 7). Hierarchical systems are organized around interrelated subsystems, with discriminable capacities. Simon thought that, as a matter of fact, not all systems exhibit interesting forms of hierarchy, but many – perhaps most – are hierarchically organized. Among those that do exhibit some degree of hierarchical order, the key feature concerns the relative strength of interaction *among* as opposed to that *within* subsystems. The clear and straightforward thought is that as interaction among subsystems increases, the significance of interaction within subsystems decreases, and conversely.[1] So in relatively simple hierarchies, there will be a relatively high strength of interaction within subsystems as compared with the interaction among subsystems. These are systems that are at least nearly decomposable (Simon, 1969; Bechtel & Richardson, 1993). Here is what Simon says:

> At least some kinds of hierarchic systems can be approximated successfully as nearly decomposable systems: . . . (1) in a nearly decomposable system the short-run behavior of each of the component subsystems is approximately independent of the short-run behavior of the other components; (2) in the long run the behavior of any of the components depends in only an aggregative way on the behavior of the other components. (1969, p. 210)

One clear thought is that nature is organized in terms of wholes and parts, mereologically. The other is that there is a kind of order to this whole–part ordering. Organization could be hierarchical without being decomposable, but when it is decomposable, or nearly so, there is a very natural kind of order. In the decomposable (or nearly decomposable) cases, there is a natural projection from component behavior to systemic behavior, both in the short and long term. The behaviors of parts are then relatively independent of the behaviors of other parts, on both longer and shorter time scales. In other cases this is not so; in these non-decomposable cases, the behavior of one component depends critically on the behavior of others. This is reasonably considered a case of "emergence," though this is emergence without magic (cf. Boogerd et al., 2005). In either case, we have a hierarchical organization.

If our theories are to reflect the structure we find in nature they too must be hierarchically organized. There are multiple levels of organization in the world, however exactly we individuate these levels. Social groups consist of individuals. Individual organisms consist of a variety of organ systems, and those in turn of organs. These organs consist of specialized cells. The cells also have parts, and those parts have parts. Cells include not only enzymes and genes, but organelles and ions. This sort of decomposition can be continued to simpler and simpler entities. The order can be reversed, at least in principle. Assuming some decomposition of wholes into constituent parts, it is tempting to think all the explanatory work can be done at the lower level. If we understand the behavior of entities at any one level, then the behavior of more complex objects consisting of the simpler entities must be a function of the behavior of the simpler constituents, given the context. In some cases, we are in fact able to determine the behavior of complexes computationally, given the behavior of parts. In some cases, we cannot, if only because the computational problems are overwhelming. Many body problems in physics are well-known cases. Genetic interaction is also illuminating for illustrating how daunting complexity becomes. Explanations of relatively "simple" genetic systems very quickly become computationally intractable as there are more interactions among the genes, more differences among alleles, and as their effects on the phenotype become more varied. Insofar as we are interested in tracking genetic changes evolutionarily, it is crucial to know not just the major effects a genetic variant will have, but the epistatic and pleiotropic effects of those genes. Despite such problems, it is hard *not* to think that if entities at higher levels are composed entirely of entities described at lower levels, and if the behavior of these simpler entities can be adequately captured by theories tailored for those lower levels, then higher order behavior must depend only on the behavior of the entities at the lower level; and likewise it is hard not to think that a theory adequate at the lower level should, in principle, be adequate for higher levels as well. The problems posed by many bodies and by complex interactions, then, seem to be simply the consequences of limitations on our cognitive powers. Were we to relax these limitations, or if we could, the solutions might become obvious. So if we begin with enzymes, and we know their kinetic properties within the cell, then we might, in principle at least, predict the pathways that emerge. If we know the boundary conditions – including the available substrates – then we can, in principle at least, "compute" the pathways. This looks like reductionism at work. I accept that it is hard not to think this way, but I also think it is wrong to think this is so (see Stephan & Richardson, 2006; Wimsatt, 1976).

2. Reduction in the *Structures*

The two *Structures* promoted a vision of reduction understood in terms of deductive logical structures, conceived independently of the limits imposed by social and individual cognitive limitations. The *Structures* saw reduction as an explanatory relation between theories, where the theories are understood as formal structures and explanation is a deductive relation. The goal of reduction is to promote parsimony in two dimensions: first, in basic ontological commitments; and, second, in basic explanatory principles. The result would be a hierarchy of theories which is unified by a theory at the lowest level, to which the laws and regularities at higher levels can be "reduced," even if only in principle. So understood, reduction promotes what Paul Oppenheim and Hilary Putnam called the "Unity of Science." We begin with a theory, including a set of laws at a higher level, or perhaps a set of phenomenological regularities characterizing the behavior of entities at a higher level. We also begin with an articulated theory at the lower level. As it is often understood, attaining unification between two theories depends on three conditions. This was, of course, canonically described in Nagel's *Structure*, though the same picture is equally part of Kuhn's *Structure*.

(1) **Compositional identities** must be advanced which specify what entities within the domain of the reduced (or secondary) science consist of in terms of the domain of the reducing (or primary) science;
(2) **Type identities** must be advanced which identify properties of reduced entities with properties of reducing entities; and
(3) **Explanations** must be available within the primary science for the laws or the phenomenological regularities of the secondary theory (perhaps as corrected in light of the primary theory).

The third requirement – which Oppenheim and Putnam term the "Unity of Laws" – is taken to presuppose the fulfillment of the first and second – which they term the "Unity of Language" (1958, pp. 3–4). Just as we must be able to translate a foreigner's language before we can assess the truth of what he says, so, it seems, the laws of the secondary science must be reformulated before they can be explained in terms of the primary science. This dependency has had the effect of focusing discussion on the unity of language, and in particular the compositional identities and bridge laws, which reduction, so understood, requires. That consequence is unfortunate insofar as it has led to neglecting the substantive restrictions imposed by the third condition: merely translating between levels hardly guarantees reducibility since that does not guarantee the sufficiency of lower-level principles. Nagel's *Structure* captured the idea this way:

> a reduction is effected when the experimental laws of the secondary science (if it has an adequate theory, its theory as well) are shown to be the logical consequences of the theoretical assumptions (inclusive of the coordinating definitions) of the primary science. (1961, p. 352)

The explanatory step, naturally, depends on "bridge laws" such as (1) and (2) connecting the theories. Reduction is thus structurally defined. Theories are conceived of as axiomatic systems. Explanation becomes derivation within an axiomatic system. Reductionism is committed to the idea that the laws which characterize higher-level theories, or at least the most adequate higher-level theories, are explicable on the basis of lower-level theories. Accordingly, theories are typically construed as deductively organized systems, and laws are taken to be unrestricted universal generalizations.

Within what Kuhn calls "normal" science, the picture is not substantially different from that offered in Nagel's *Structure*. As Kuhn says, normal science "is an object for further articulation and specification under new or more stringent conditions" (1962, p. 23). This is the articulation of a paradigm. This can include fixing the value of physical constants, or the parameters for a model. One of Kuhn's examples is working out, and then refining, the value of the gravitational constant after Newton. This articulation can involve determining the "shape" of empirical laws. Boyle's work on the gas laws is a clear example of such articulation; it is a case to which I'll return. Kuhn calls this

"Baconian science," meaning by that areas that come to prominence because they insist on quantitative data (cf. Kuhn, 1977, ch. 8).

The articulation of scientific models may involve no more than refining quantitative models to fit particular domains. It may also involve developing applications to a new area. So, for example, the elaboration of thermodynamics toward the end of the nineteenth century required a variety of conceptual innovations. I'll return to this case in what follows. Kuhn famously has his more radical moments, when simple articulation shifts toward a picture that is no longer cumulative. Baconian science yields to Copernican. He says, for example,

> a new theory, however special its range of application, is seldom or never just an increment to what is already known. Its assimilation requires the reconstruction of prior theory and the re-evaluation of prior fact, an intrinsically revolutionary process that is seldom completed by a single man and never overnight. (1962, p. 7)

The ambiguities in Kuhn's views are well known. I want merely to underscore the thought that his picture of the progress of normal science is not all that different from the vision offered as canonical in Nagel's *Structure*. Even incommensurability – the product of revolutions when theories change – is a consequence of that very picture of the structure of theories. The lack of comparability is a consequence of substantially changing the "shape" of the theory rather than tinkering with the details. There is no simple transformation turning a duck into a rabbit.

Controversies surrounding the standard rendering of explanatory reduction by Nagel led to a number of interesting and important modifications and extensions of the classical model (cf. Schaffner, 1967; Hooker, 1981; Churchland & Churchland, 1990; Bickle, 1998). I am convinced that the consequences for the understanding of reduction are less dramatic than their authors imagine. They do manage the formal problems, but I think don't manage the larger problems with the overall pattern. Jaegwon Kim is a natural example of someone who thinks that the problems can be handled by tinkering with the details in structural models. Others have entertained similar thoughts. Changes in the models of reduction in the ensuing decades were, among other things, supposed to ameliorate the failings of the narrower structural rendering, such as the resistance of multiple realization. I also have pressed that there are "easy" fixes to some of the technical problems (Richardson, 1979). Such "fixes" are, particularly, designed to accommodate the inability of a structural rendering of theories to deal with scientific change, and the complexities associated with interlevel relations. As is reflected in the last quote from Kuhn, with a change in any of the fundamental principles of a theory, there will be changes in the interpretation of all the subsidiary principles. Given a more or less holistic rendering of theories (or disciplines), even a change in more minor principles will entail some change elsewhere. Any change leads to incommensurability. If the interpretation cannot flow from the bottom up, but depends crucially on the theoretical context – a point both Kuhn and Feyerabend (1962) embraced – then the interpretation must mean there is some change of content. Again, this is incommensurability.

In more recent treatments, this dialectic is clear enough. Kenneth Schaffner's elaborate development of a structural model maintains the core idea that unifies the *Structures*, within a formalist framework. This was originally advanced in a definitive article aimed at incorporating the various approaches to reduction that were then in place (Schaffner, 1967). He has since (1993) elaborated a very comprehensive treatment that maintains the structural line, in broad outlines, and that dramatically extends its scope. It is not so much to my point to criticize that view here, or to endorse it, so much as to notice the power and scope that is available to structural renderings (see Richardson & Stephan, 2006). Schaffner describes a two-tiered model (Figure 6.1). Between levels, we have explanatory relationships understood in a classical way. A lower-level theory (say, biochemistry) is used to explain the behavior of something captured at a higher level (say, the immune response of organisms). Among theories at the same level, we expect at least some sort of analogy, or structural similarity. So reduction between levels is still an explanatory (deductive) relation between levels, though there is no requirement of an exact fit between what gets explained and the phenomena we started with.[2] John Bickle adopts a modestly different approach toward capturing the phenomena, but it is one that has a similar effect. In *Psychoneural Reduction* (1998), Bickle acknowledges, as does Schaffner, that there can be failures of match between reduced and reducing theories. Drawing on work within philosophy of science, and especially European "Structuralist" work on the semantic view

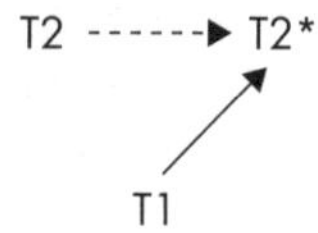

Figure 6.1 Schaffner's two-tiered model. Time is represented on the horizontal axis, so that a reduced theory (T2) is succeeded by an "improved" alternative based on a theory (T2*) at a lower level of organization (T1). T2* improves on T2, and is deductively explained by T1. The schema is derived from Schaffner (1967), and was elaborated in Schaffner (1993).

of theories by Balzer, Moulines, and Sneed (1987),[3] Bickle wants to convince us that reduction is alive and well even within cognitive neuroscience. Like Schaffner, Bickle claims that what is derived from the lower level is a conclusion that is at best *analogous* to what we began with. Like Schaffner, Bickle retains the idea that what is essential to successful reductions, as opposed to eliminations, is retention of structure.

Within a Structural approach, explicability of the laws of the higher-level theory in terms of lower-level theories is sufficient for the most central commitment to the dispensability "in principle" of higher-level theories in favor of lower-level theories, and would therefore be sufficient to promote the twin commitments to theoretical and ontological simplicity that are constitutive of philosophical reductionism. If everything that can be explained is explicable using only the reducing theory, then the reduced theory would add nothing of substance. Reduction so understood licenses elimination. This raises a variety of interesting issues. Centrally, it is clear enough that even if X is composed entirely of Y, it doesn't follow that beginning with Y will make it possible to explain X. It is one thing to effect a decomposition. It is still another to enable a construction from the resources at the lower level alone. Descendants of the *Structures* assume that this is something of a distraction; with enough computational power, they think, the construction must be possible. Within a naturalistic philosophy of science, I think the Structuralist response is problematic at best. Science is, after all, conducted within the constraints of bounded rationality (cf. Bechtel & Richardson, 1993; Stephan & Richardson, 2006). Furthermore, I think that the imposition of higher-level constraints is crucial in the development of explanatory models

– mechanistic models – and so am convinced that construction from the bottom up is illusory. I'll try to illustrate these points briefly in the following section, using a case that would surely count as favorable to the account of the *Structures.*

3. Thermodynamics and Statistical Mechanics

The paradigm case within philosophy of science for intertheoretic reduction is certainly the reduction of classical thermodynamics to statistical mechanics (Nagel, 1961; Bickle, 1998). The example has been widely discussed, and forms the centerpiece for discussions of reduction (see Sklar, 1993 for a comprehensive discussion). It is broadly regarded as a theoretical reduction, pitched to different levels of organization. It is thought of as what Robert Causey (1977) called a *microreduction*. Let's start with a relatively standard portrayal. I certainly do not mean to suggest that the standard portrayal is correct, though it certainly resonates with some of the science. I am convinced that in many ways it is mistaken. As Structuralists deploy it, it is intended as a kind of "rational reconstruction," so much of the nuance I think is important might be considered noise by proponents. What immediately follows is intended to recount the Structuralist rendering of the case. I will modify it shortly.

The reduction of thermodynamics to statistical mechanics was consolidated toward the end of the nineteenth century in the work of Maxwell and Boltzmann. It received a canonical description by Josiah Gibbs in the early twentieth century. Thermodynamics embodied a number of phenomenological laws describing the behavior of gases. Among the most important were Boyle's law and the law of Charles and Gay-Lussac. Boyle's law says that, at least as a first approximation, the product of the pressure (P) and volume (V) for a sample of gas is constant at a constant temperature; that is, with a constant temperature, pressure and volume vary inversely so that increasing one proportionally decreases the other. The law of Charles and Gay-Lussac says essentially that any increase in volume as a function of increasing temperature (T) is constant for a gas held at a constant pressure; that is, with a constant pressure, changes in volume and temperature vary directly with one another. The combined law is a straightforward consequence of these two principles. It says that, again to a first approximation, the product of pressure and volume is a constant function of temperature. In a textbook form, the Boyle–Charles law is this, with k a gas specific constant:

$$PV = kT$$

Within statistical mechanics, we can derive an analogous result. Under a variety of simplifying assumptions, it is possible to show that the product of pressure and volume is a constant function of the aggregate kinetic energy (e_t) of a gas. This is known as the Bernoulli formula:

$$PV = (2/3)e_t$$

The results bear obvious similarities so expressed. This motivates the conclusion that the thermal energy contained in a quantity of gas is simply the aggregate kinetic energy of the particles constituting that gas, and that changes in the temperature of the gas are equivalent to changes in the aggregate kinetic energy of these particles. The Bernoulli formula allows us to explain what is captured by the Boyle–Charles law, and even to explain why the Boyle–Charles law holds as nearly as it does. It is also possible to explain why the Boyle–Charles law fails at extremes of temperature and density, though not in as direct a way as we can explain the law itself. The kinetic theory of gases is thus capable of explaining much of what is explained by classical thermodynamics, and can explain the behavior of gases with greater precision. Classical thermodynamics is therefore held to reduce to statistical mechanics.

The actual history is more interesting and more complicated than the structural rendering suggests, though what I say here can be only suggestive. One place to begin is in the eighteenth century, with a formulation of a theory of heat by Joseph Black (1728–1799). Black's theory was designed to embrace a variety of simple thermal phenomena, and to provide a means for implementing and controlling engineering problems. The crucial phenomenon for which Black's theory was designed is thermal equilibrium; that is, the simple fact that systems tend to attain a state in which the temperature is homogeneously distributed. In observing phenomena connected with thermal equilibrium, it was noted that the progress toward the attainment of thermal equilibrium suggested (1) that the process could usefully be conceptualized as a flow of heat from the warmer body to a colder one with which it maintained contact – or from one part of a body to another cooler part; and (2) that the rate at which this process occurred depends on the specific nature of the bodies involved. Black supposed that the process could be usefully analogized to a fluid transfer from one source to another, and that thermal equilibrium could be understood as analogous to equalization of the levels of water in connected containers. Accordingly, he supposed that progress toward thermal equilibrium involved the transfer of a weightless fluid capable of permeating material substances. This fluid was later termed "caloric." The rate at which the progress toward thermal equilibrium in heating a homogeneous substance occurred was, in turn, seen to a constant function of its mass, where the constant is dependent upon the specific substance involved. This parameter, the "specific heat" of the substance (that is, the amount of heat necessary to raise a unit mass of that substance by 1°), served as the main explanatory parameter in Black's theory.

Black's theory enjoyed widespread success, and served as an empirically plausible and adequate theory for explaining the phenomena associated with thermal equilibrium and thermal changes so long as the temperature intervals involved were relatively small. Yet it became clear by the beginning of the nineteenth century that the "caloric" theory could not allow for a theory of heat that would unify the treatment of the dynamics of thermal equilibrium with heat in other forms; moreover, it was clear that such a unified treatment was mandatory, given the interconversion of heat in its various forms. Thus, the phenomena associated with the mechanical production of heat (for example, in friction) and radiant heat (which by then was recognized to be strongly analogous to light conceived in accordance with the then-prevailing wave theory) came to occupy a central role, as anomalies. Any theory, such as Black's, which precluded a unified account was regarded a problematic. Thermodynamics in a recognizably modern form was born.[4] In the newly emerging theory, it was useful to begin with the simplest case. That, of course, is matter in a gaseous state. Since gases are relatively unstructured, they exhibit simpler dynamics. It is in this very limited but theoretically important context that Boyle's law and the law of Charles and Gay-Lussac were significant. The combined law is, as is and was well known, strictly limited in scope. The regularities it expresses are maintained only for a few "ideal" gases at low densities and moderate temperatures. At extreme temperatures or high densities, the law breaks down completely. This is thus one among the "ideal gas laws," so termed as descriptions of what "ideal" or "perfect" gases would do.[5]

Though the Boyle–Charles law describes and is geared to explain one of several macro-properties of low density and medium temperature gases, it has assumed an important role in the unification of thermodynamics and kinetic theory. It is in fact one of the central postulates which a unified theory needs to explain, just as the unified theory needs to explain the well-known deviations from the law. This yields the standard structural rendering of the case, beginning with but three postulates:

(1) Compositional identity: Gases can be identified with aggregates of particles which lack any internal structure.

Bernoulli had suggested that a gas should be thought of as an aggregate of molecules in motion late in the eighteenth century, and clearly connected this model with what was later Boyle's law. It was not until the middle of the nineteenth century that the kinetic theory began to gain prominence, largely under the influence of Maxwell and Boltzmann.[6]

(2) Constituents: The particles constituting a gas are of a constant size, being vanishingly small in comparison with the distance between them.

In the simplest treatment, the molecular constituents are taken to be point masses. Of course, no one thought that that was a literally true description. The principle is motivated by the fact that it promotes computational efficiency, rather than any merits of realism.

(3) Thermal energy: The thermal energy of a gas is to be identified with the aggregate kinetic energy of the particles constituting the gas, and an increase/decrease in the temperature of the gas is due to an increase/decrease in the aggregate kinetic energy of these particles.

This is the key "bridge law." In the structural representation of the classical case above, this last principle was the key to the reduction. With these assumptions, it is a relatively trivial matter to explain the equation of state, on the assumption that Newton's laws of motion apply unchanged at the atomic level (a false assumption, but one that, in the time period we are dealing with, no one had reason to doubt). It is a matter simply of calculating the average force exerted on the sides of a container by a gas with a known number **N** of molecules of determinate mass **m**, which in turn yields the Bernoulli formula.

So far the story matches reasonably well with the vision of the *Structures*, especially with the modifications its heirs have introduced. The isomorphism between the Boyle–Charles Law and the Bernoulli formula at least illustrates the idea that there can be formal similarity of results. In other cases, when formal similarity fails, the expected result will be the elimination of the reduced theory in favor of its more precise and successful counterpart. The interpretation transposes in a straightforward way to the paradigm case:

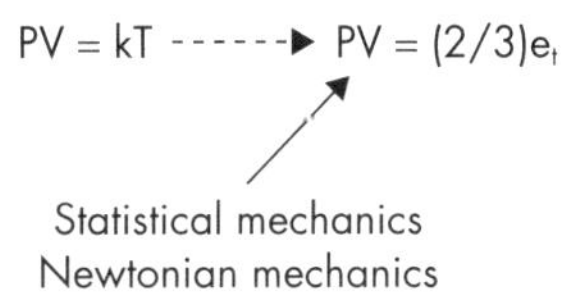

Figure 6.2 A two-tiered representation of thermodynamic reduction. The schema follows that represented in Figure 6.1. What is derived from statistical mechanics and Newtonian mechanics is a principle analogous to the Boyle–Charles law

Analogy, indicated on the horizontal, is of course an unsatisfying cornerstone for a theory of reduction; it would certainly be rejected by those that wanted the clarity of a formal analysis. Still, it has a number of virtues, among them that it allows reduction to be a matter of degree. Theories may be more or less conserved, depending on whether they map more or less well onto their successors.

I think the more historically realistic account suggests that, contrary to the model of the *Structures*, the motive force for the unification of kinetic theory and thermodynamics was not a desire to explain or derive the "laws" of thermodynamics from Newtonian physics; ontological economy is an even more remote issue. One motivation seems to be a desire to *situate* thermodynamics within a Newtonian theory. This was accomplished by showing that the gas laws are *consistent* with a Newtonian vision, in which molecules are conceived as Newtonian point masses. Demonstrating consistency falls far short of *deriving* the gas laws from the dynamics of particles. In point of fact, no one attempted such a derivation in even one case. Such an attempt would be computationally intractable. Another motivation was the need to explain the interconversion of thermal energy with other forms, and this could not be done without a serious recasting of the problem. There is, significantly, a degree of bidirectional exchange between theories at the two levels; in particular, the model of gas structure which resulted was designed to systematize the phenomena described by Boyle, Gay-Lussac, and others. Nonetheless, there is an asymmetry, since the physical principles brought to bear – in particular, the application of Newtonian principles to atomic phenomena – are simply incorporated unchanged. Perhaps most fundamentally, the explanation of the gas laws is irreducibly statistical. It is a theory of aggregate behavior, and not a derivation from Newtonian first principles.

4. Rethinking Reduction as a Dynamic Relation

Such cases suggest the following picture. A reductionistic strategy is one that seeks explanations for selected higher-level phenomena in terms of mechanisms described and motivated at the lower level. The phenomena to be explained may be simple patterns of behavior, or anomalies, or both. Thus, in the case of thermodynamics, interconversion among the forms of heat is a natural outcome since there is a common currency in kinetic energy. Likewise, in chemistry, the behavior of isomers was given an explanation in terms of atomic structure. There need not be any systematic retention of the theoretical categories which dominate theorizing, in the initial theory, and, characteristically, there need not be a retention of explanatory tools which are motivated by the demands within the antecedent theory. Black's account of thermal dynamics, for example, were replaced by models that were more adequate, and not captured at all by later theories; that is, Black's theory was displaced rather than retained.

It is this asymmetry in theories which gives some credibility to eliminationist theories such as those defended by Feyerabend (1962) and, more recently, by Paul Churchland (1979), Patricia Churchland (1986) and John Bickle (1998). Nonetheless, if I am right, neither economy nor elimination is the goal of "reduction," and to emphasize elimination in these interlevel cases requires doing considerable violence to the history of the sciences: there is a significant continuity in the scientific concerns, and higher-level theories are not eliminated but recast and integrated. Eliminative pictures are, in short, driven by a structural vision of reduction, a vision which does not comport with the history of science. The shift I recommend is from thinking of theories, and of theory elimination, toward explanatory goals, and a coordinate shift toward recognizing the explanatory benefits sometimes gained by shifting to lower levels, or to higher levels. I'll give some reason to underscore this idea presently.

If the motivating force for adopting a reductionistic research strategy is explanatory adequacy, we can understand some other typical features of reductionistic science. It is for this reason that anomalies, whether experimental or theoretical, play such crucial roles in reductionistic theorizing: experimental anomalies (such as the behavior of isomers in molecular bonding) suggest a failure of higher-level models to explain the phenomena within their domain, and theoretical anomalies (such as the interconversion of thermal energy with mechanical) suggest that the models fail to unify significantly similar domains. I think Wimsatt drew *exactly* the right moral in thinking about cases of this sort:

> When a macro-regularity has relatively few exceptions, redescribing a phenomenon that meets the macro-regularity in terms of an exact micro-regularity provides no (or negligibly) further explanation. All (or most) of the explanatory power of the lower level description is "screened off" . . . by the success of the macro-regularity. The situation is different however for cases which are anomalies for or exceptions to the upper level regularities. Since an anomaly does not meet the macro-regularity, the macro-regularity cannot "screen off" the micro-level variables. If the class of macro-level cases within which exceptions occur is significantly non-homogeneous when described in micro-level terms, then going to a lower-level description can be significantly explanatory. (1974, p. 690)

Even if we focus on the paradigmatic reductionist cases, the drive is *not* toward lower-level explanatory sufficiency, but is geared toward gaining *overall* explanatory adequacy. The problems for which the lower-level models and theories are designed are but one element in that quest. There will be an emphasis on explaining selected upper-level phenomena and limited upper-level regularities in terms of mechanisms framed at the lower level, but no demand to explain all the upper-level phenomena in lower-level terms. To take another example, the explanation of genetic replication and a variety of other phenomena such as dominance and linkage were central problems which lay beyond the scope of Mendelian genetics. Classical genetics could see the phenomena, but could not explain them. The former problem was resolved by molecular genetics after Watson and Crick. The latter are now largely understood, as consequences of the biochemistry of gene action and of molecular structure. What was successfully explained by Mendelian genetics did not *need* explaining in terms of molecular genetics (see Kitcher, 1984, pp. 358 ff.). Hence, reductionistic strategies do not mandate the attempt to eliminate higher-level theories, or to "displace" them in favor of theories and models framed solely in terms appropriate to the lower level. A reductionistic strategy thus suggests a process that is more akin to the construction of a unified theory with causal parameters at multiple levels, employing lower-level mechanisms to explain significant upper-level phenomena, than to anything describable within a classical model of reduction (cf. Darden & Maull, 1977; Maull, 1977; Darden, 1980). The key thought is that it is not the structure that illuminates reduction, but the dynamics, and the dynamics are driven by explanatory rather than metaphysical needs.

A similar point underwrites reductionist failures. In some cases, we find explanatory anomalies that get explained by shifting "up" a level in terms of theories or models; that is, we explain a phenomenon by embedding it in a broader context. I've discussed such cases elsewhere in the setting of work on ontogeny, or development. In cases such as differentiation of structures in *Drosophila*, there are dramatic and well-understood influences outside the organism that are critical in explaining the developing organism. Polarity in the developing embryo depends critically on maternal influences, and without those influences there is no structure on which to build developmental structures (cf. Richardson, 1998). William Wimsatt has developed this theme extensively in conjunction with what he calls "developmental entrenchment." So, sometimes, shifts to higher levels can facilitate explanatory fecundity. As with shifts to lower levels, the goal is explanatory, and the outcome, if successful, is enhanced overall explanatory power.

5. Resolving the *Structures*

One core idea shared by the two *Structures* is that reduction is a matter of explanation – an explanatory and deductive relation between theories, models, or sciences in which one theory, model, or science captures or explains the other, with some gain in precision or scope. Reduction, so understood, may involve theories at the same level of organization or at different levels of organization; and reduction may involve one theory explaining another theory or explaining only the empirical phenomena captured by the reduced theory. On the account of the *Structures*, all these cases have the same structure. The range of cases are given a uniform analysis in terms of the structure of theories and deductive relations. Thomas Nickles (1973), William Wimsatt (1976), and Robert McCauley (1986, 1996) have each insisted that it is a mistake to conflate reduction as it occurs *within* and *between* levels of organization. I am very much in sympathy with that general point (cf. Richardson, 2002). My focus here has been on cases that integrate multiple levels of organization rather than on cases involving theories at the same level of organization. I have urged that we should focus on the dynamics rather than the structures. Though the description I've offered of the dynamics is geared toward interlevel cases, I think the shift toward dynamics driven by explanatory demands can illuminate both classes of cases.

The two classes of cases evidently do exhibit importantly different dynamics. I made a good deal, in the first two sections, of the fact that the two *Structures* focus on different paradigms. Kuhn focuses on theories that are broadly competitors, and what we see historically is that the differences tend to exclude one in favor of the other. In the end, Ptolemy's astronomy is not conserved through the Copernican revolution. A system of concentric spheres, carrying planets on their circuits, is lost, even though Copernicus could scarcely do better from an empirical standpoint. These are intralevel cases. They are typically *not* conservative. The antecedent theories are not retained, even to a first approximation. The phenomena are, at least in many cases, retained, though even that is not universally true. The explanations and theories are not conserved. By way of contrast, Nagel focuses on theories that are not actually competitors, and are naturally thought of as occupying different levels of organization. He thinks of statistical mechanics as pitched toward a lower level, and providing the explanation – a deductive explanation – for the gas laws.[7] Assuming that is true, it would be surprising to find that, for example, the relations between pressure and volume are not what we antecedently expect on the basis of the macroscopic laws. More basic theories should not contradict, much less displace, less basic and more mature theories, since they explain those theories. The more recent accounts of theory reduction I've discussed did, of course, allow for "corrections" from the lower level, but at least in the more conservative cases the upper-level laws held at least to a first approximation. Indeed, a more "basic" theory that did not explain the phenomena explained at the higher level would be false. To return to the case discussed in some detail in Section 3, that is why it was important to show that an atomist view was consistent with thermodynamics. These are interlevel cases.

Once we attend to the disparity between inter- and intralevel cases, I think several morals emerge. I cannot do more than indicate the broad directions this takes us. First, as I've suggested, interlevel cases are naturally more conservative. That doesn't mean that higher levels survive unscathed. Neither do lower levels. The process that we typically see is an interweaving of levels, with constraints from multiple levels at once. These constraints may come from both higher and lower levels of organization. As Simon recognized, the processes often operate at different temporal scales. It would be surprising to see someone announce that, having looked in detail at proteins and enzymes, and examined their interactions with cellular substrates, there are after all no biochemical pathways or no ribozomes. In fact, we should regard the conclusion as simply false. After all, it is false. Biochemical pathways are as real as proteins, and feature importantly in explanations of cell physiology. These pathways are indeed constituted by the activities of enzymes, and describe the dynamics of systems; but that does not compromise the explanatory status of pathways or their reality. Enzymes operate at one temporal scale. Pathways are constituted at a longer time scale, often involving significant regulation.

The uptake of substances from the environment is typically much slower than the processes that consume them within the cell; and regulatory loops often involve many transformations. On the other side, ribozomes are important biological structures, implicated in protein synthesis. The fact that much of protein metabolism can be understood without shifting to a level that involves protein synthesis and gene expression does not compromise the reality of entities understood at the lower level. The fact is that the physiological reactions among enzymes and substrates operate at a much faster time scale than does gene expression and protein synthesis.

Second, while it is often important in intralevel cases to show how some range of phenomena can be captured by a competitor, that is not the typical practice in interlevel cases. In the previous section, I observed that, though research in transmission genetics within the Morgan school could, and did, reveal, phenomena such as incomplete dominance, they could not explain it.[8] That required a shift to a lower level, as did the understanding of more complex pathways and interdependencies (cf. Bechtel & Richardson, 1993). Wimsatt suggests the key rule is to explain everything that happens and nothing that doesn't. We should add that it is not necessary to explain anything twice. As he says, there is not explanatory gain in explaining something twice. Wimsatt's rule is compatible with turning to lower levels in order to explain things left unexplained at the higher level. It is, of course, also compatible with turning to higher levels in order to explain things left unexplained at a lower level. Third, a dynamic model, in which we attend not just to theories, but to experimental methods, to the salient phenomena, to the development and modification of models, and to the context in which research is conducted, is likely to be more revealing concerning the differences among the cases. Some phenomena live at a time scale that requires an explanation at one level of organization. So if we want to explain how *E. coli* shift toward the consumption of lactose as simpler sugars are consumed, it is important to recognize that this requires the synthesis of digestive enzymes that can consume lactose. If we want to explain how *E. coli* evolved this capacity, we need to look toward very different mechanisms, concerned with genetic mutation and selection. Or to use a very different example, if we want to explain the patterns of adaptation in a species of bird on a volcanic island, that will depend on the variations available, the selection pressures, and the population structure.

Explaining extinction in the face of an environmental catastrophe such as the introduction of an exotic predator operates on a radically different time scale. There is simply no time for adaptation, and the result is collapse.

Such a focus requires that we shift toward thinking about the dynamics of theories, toward the development of research programs, rather than focusing on the structures. It also requires that our explanatory theories reflect the differences in the mechanisms. The vision fits comfortably with Simon's, recognizing that nature is hierarchically structured, with processes operating at different time scales. We should expect that our science reflect natural divisions, and our sciences respect the differences in nature.

Acknowledgments

I have been fortunate to discuss this material with a number of colleagues, including John Bickle, Fred Boogerd, Frank Bruggeman, Peggy DesAutels, Lawrence Jost, W. E. Morris, Thomas Polger, Robert Skipper, and Achim Stephan. I've been influenced in many ways by William Bechtel, Robert McCauley, and William Wimsatt. In addition, I've profited from discussions of the article at the Free University in Amsterdam, the University of Osnabrück, and the University of Cincinnati. The Taft Faculty Committee at the University of Cincinnati has generously supported the work.

Notes

1 Discussions concerning "modularity" are affected by this issue. Some claim there is "massive modularity," intending by this that the mind/brain is organized in a way that achieves relative independence for some parts; others argue that there is more "plasticity," evidently intending that there is more modifiability in brain structures. I think the controversy is spurious. The real question, which few address, is the extent of local control as opposed to global control over systemic behavior.

2 Cliff Hooker (1981) also develops a model, in considerable detail, with the same fundamental structure.

3 The German Structuralists embraced a vision of theories that has similarities to those of Kuhn and Nagel. Bickle followed them in emphasizing semantic rather than syntactic structures. Unlike Bickle, these German

Structuralists emphasized the importance of scientific change, relying on a largely Lakatosian scheme, noting the conservation of the core assumptions. This is an important departure, despite the shared assumptions about the structure of theories.

4 In one of its more natural formulations, the first law of thermodynamics – the law of conservation of energy – states that for any thermodynamic system, a change in its internal energy is a function of the heat gained or lost and of the mechanical work done. It thus implicitly presupposes the interconversion between mechanical energy and thermal energy.

5 A second principle applying to the "ideal" gases was confirmed and elaborated by Joule in collaboration with Kelvin in the latter part of the nineteenth century. This principle, generally called "Joule's Law," states that the energy in an ideal gas at a constant temperature is independent of any change in volume. This principle connects, on the one hand, with the theory of specific heats and, on the other hand, with a kinetic theory of gases. Like Boyle's law and the law of Charles and Gay-Lussac, when the density or pressure of the gas is high, the constant ratio of temperature and internal energy breaks down.

6 Buchdahl's *The Concepts of Classical Thermodynamics* (1966) is a good treatment of the history of thermodynamics; and Brush's *Statistical Physics and the Atomic Theory of Matter* (1983) is a brief treatment of the kinetic theory.

7 I'm not altogether happy with that picture, but that has not occupied me here. The problem I see is that, in fact, the "reducing" theory is a theory of the statistical behavior of aggregates, and is not constructive. It can be understood as a theory pitched toward the same level of organization as the theory of gases.

8 One model that explained dominance in terms of *absence* could explain simple dominance; but to explain partial dominance depended on understanding enzymatic activities.

References

Balzer, W., Moulines, C. U., & Sneed, J. (1987). *An Architectonic for Science.* Dordrecht: Reidel.

Bechtel, W., & Richardson, R. C. (1993). *Discovering Complexity: Decomposition and Localization as Strategies in Scientific Research.* Princeton: Princeton University Press.

Bickle, J. (1998). *Psychoneural Reduction: The New Wave.* Cambridge, MA: MIT Press.

Boogerd, F. C., Bruggeman, F. J., Richardson, R. C., Stephan, A., & Westerhoff, H. V. (2005). Emergence and its place in nature: A case study of biochemical networks. *Synthese*, *145*, 131–164.

Brush, S. (1983). *Statistical Physics and the Atomic Theory of Matter*. Princeton: Princeton University Press.

Buchdahl, H. (1966). *The Concepts of Classical Thermodynamics*. Cambridge: Cambridge University Press.

Causey, R. (1977). *Unity of Science*. Dordrecht: Reidel.

Churchland, P. M. (1979). *Scientific Realism and the Plasticity of Mind*. Cambridge: Cambridge University Press.

Churchland, P. M., & Churchland, P. S. (1990). Intertheoretic reduction: A neuroscientist's field guide. *Seminars in the Neurosciences*, *2*, 249–256.

Churchland, P. S. (1986). *Neurophilosophy: Toward a Unified Theory of the Mind–Brain*. Cambridge, MA: MIT Press.

Darden, L. (1980). Theory construction in genetics. In T. Nickles (Ed.), *Scientific Discovery: Case Studies* (pp. 151–170). Dordrecht: Reidel.

Darden, L., & Maull, N. (1977). Interfield theories. *Philosophy of Science*, *43*, 44–64.

Feyerabend, P. (1962). Explanation, reduction and empiricism. In H. Feigl & G. Maxwell (Eds.), *Minnesota Studies in the Philosophy of Science* (Vol. III, pp. 28–97). Minneapolis: University of Minnesota Press.

Hooker, C. A. (1981). Towards a general theory of reduction. Part I: Historical and scientific setting. Part II: Identity in reduction. Part III: Cross-categorical reduction. *Dialogue*, *20*, 38–59, 201–236, 496–529.

Kitcher, P. (1984). 1953 and all that: A tale of two sciences. *Philosophical Review*, *93*, 335–373. Reprinted in P. Kitcher, *In Mendel's Mirror* (pp. 3–30). New York: Oxford University Press, 2003.

Kuhn, T. S. (1962). *The Structure of Scientific Revolutions*. Chicago: University of Chicago Press.

Kuhn, T. S. (1977). *The Essential Tension*. Chicago: University of Chicago Press.

Maull, N. (1977). Unifying science without reduction. *Studies in the History and Philosophy of Science*, *8*, 143–162.

McCauley, R. N. (1986). Intertheoretic relations and the future of psychology. *Philosophy of Science*, *53*, 179–199.

McCauley, R. N. (1996). Explanatory pluralism and the co-evolution of theories in science. In R. N. McCauley (Ed.), *The Churchlands and Their Critics* (pp. 17–47). Oxford: Blackwell.

Nagel, E. (1961). *The Structure of Science: Problems in the Logic of Scientific Explanation*. New York: Harcourt, Brace, and World.

Nickles, T. (1973). Two concepts of intertheoretic reduction. *Journal of Philosophy*, *70*, 181–201.

Oppenheim, P., & Putnam, H. (1958). Unity of science as a working hypothesis. In H. Feigl, M. Scriven, & G. Maxwell (Eds.), *Minnesota Studies in Philosophy of Science* (Vol. II, pp. 3–36). Minneapolis: University of Minnesota Press.

Richardson, R. C. (1979). Functionalism and reductionism. *Philosophy of Science*, *46*, 533–558.

Richardson, R. C. (1998). The organism in development. *Philosophy of Science*, *67*, S312–S321.

Richardson, R. C. (2002). Reduction. In *Encyclopedia of Cognitive Science*. New York: Macmillan.

Richardson, R. C., & Stephan, A. (2006). Reductionism (anti-reductionism, reductive explanation). In M. Binder, N. Hirokawa, & U. Windhorst (Eds.), *Encyclopedic Reference of Neuroscience*. Heidelberg: Springer.

Schaffner, K. F. (1967). Approaches to reduction. *Philosophy of Science*, *34*, 137–147.

Schaffner, K. F. (1993). *Discovery and Explanation in Biology and Medicine*. Chicago: University of Chicago Press.

Simon, H. (1969). *The Sciences of the Artificial* (2nd ed.). Cambridge, MA: MIT Press.

Sklar, L. (1993). *Physics and Chance: Philosophical Issues in the Foundations of Statistical Mechanics*. Cambridge: Cambridge University Press.

Stephan, A., & Richardson, R. C. (2006). Mechanisms in systems biology. In F. Boogerd & F. Bruggeman (Eds.), *Theory and Philosophy of Systems Biology*. Amsterdam: Elsevier.

Wimsatt, W. C. (1974). Reductive explanation: A functional account. In A. C. Michalos, C. A. Hooker, G. Pearce, & R. S. Cohen (Eds.), *PSA 1974* (pp. 671–710). Dordrecht: Reidel.

Wimsatt, W. C. (1976). Reductionism, levels of organization, and the mind–body problem. In G. Globus, G. Maxwell, & I. Savodnik (Eds.), *Consciousness and the Brain* (pp. 196–267). New York: Plenum.

7

REINFORCING THE THREE "R"S: REDUCTION, RECEPTION, AND REPLACEMENT

Ronald Endicott

1. Introduction

Discussions of reduction and related issues in the philosophy of science are complicated by at least two factors: *ambiguity*, or multiple concepts of reduction, and *ambition*, or the range of cases to which a given concept of reduction is thought to apply. First, regarding ambiguity, "reduction" expresses different concepts to different individuals and intellectual communities. For example, in contrast to evolutionary explanations, the biologist Theodosius Dobzhanski identifies reduction with certain Cartesian methods employed in mechanistic explanation, and he cites the mathematical treatment of automata as a case in point (1968, pp. 1–2). Yet, turning to a different intellectual community, many philosophers who are trained in classical computational psychology would not judge the mathematical treatment of automata and various complementary accounts of their mechanistic implementations to be reductionist in any substantial sense, being compatible with the autonomy of computational theory and the nonidentity of computational and physical properties.

Thus, Jerry Fodor – no friend of "reduction" in the philosophy of mind – presents the computational model as a grand synthesis of mechanism and psychological explanation (1981b, pp. 13ff.). For someone like Fodor, the concept of reduction is more closely aligned with ideas that descended from the logical positivist tradition, specifically, the absorption of one theory by another via connecting principles that express property identities between two theoretical domains. That notion is not implied by mechanistic explanation *per se*. Indeed, some philosophers claim that notions based upon mechanistic explanation provide a better alternative to the notions of reduction that have dominated philosophical discussions in the post-positivist era (Bechtel & Richardson, 1993; Machamer, Darden, & Craver, 2000).

Second, regarding ambition, after one settles upon a particular concept of reduction, and after one finishes the modest philosophical work of clarifying or refining the concept in question, there is the empirical task of determining the range of cases to which that concept of reduction applies. Here the most serious controversies arise, since virtually every existing concept of reduction applies *somewhere*, if only to marginal but well chosen cases. So the most important question is whether a suitably refined concept can apply to a satisfactory range of cases which a critical mass of scientists describe as reduction, or, more ambitiously, whether, by the discovery of new information or a reconceptualization of the old, that concept of reduction can be extended to other unanticipated cases. Here the philosopher and the scientist face the danger of being overly ambitious. There is a significant difference between ***providing an account of scientific reduction*** versus ***defending a broad philosophical vision of physicalistic monism*** whereby all theories are either reducible to or replaceable by theories in the physical sciences. For example, various psychological theories remain resistant to the concepts of reduction that developed out of the positivist tradition. Hence, theoreticians should be prepared to apply those concepts of reduction to the appropriate range of cases where the world happens to comply even if the world does not always so comply.

With these preliminaries in hand, I begin by describing a concept expressed by Kenneth Schaffner's "General Reduction-Replacement" account of scientific unification. I then turn to its popular descendant, the "New Wave" approach developed by Paul and Patricia Churchland, Clifford Hooker, and John Bickle. Both Schaffner and the New Wave interpret scientific unification very broadly in terms of a continuum of cases from theory reduction to theory replacement. This is good insofar as it goes. But I propose to expand the picture in a way that is more receptive to the role that otherwise and in other respects irreducible and irreplaceable theories play in a process of partial reduction, specifically, their token reduction. The result is a more comprehensive "Reduction-Reception-Replacement" model of scientific unification.

2. Schaffner's General Reduction-Replacement Paradigm

Schaffner's views on reduction are meant to be an improvement upon the classical position that developed within the logical positivist tradition. Consider Ernest Nagel's view of reduction, which represents the paradigm of positivist thinking on the subject. According to Nagel (1961), "reduction" meant the derivation of a theory from a more basic or inclusive theory, in a heterogeneous case where the theories do not share a common vocabulary, by means of connecting principles that link the theoretical terms of the respective theories. As this tradition developed, the connecting principles were conceived in terms of biconditional bridge laws that express cross-theoretic property identities (Sklar, 1967, pp. 120–124; Causey, 1972).

Parenthetically, Robert Richardson (1979) and John Bickle (1998, p. 120) claim that Nagel held a more liberal view, believing that the purposes of reduction could be served by weaker one-way conditionals that express mere sufficient conditions in the basic reducing theory. But the justification is based upon a footnote in Nagel which I believe has been misunderstood.[1] After mentioning Kemeny and Oppenheim's observation that connectability would guarantee derivability if the connections between the theories were biconditional in form, Nagel says: "However, the linkage between *A* [the nonbasic or 'secondary' science] and *B* [the basic or 'primary' science] is not necessarily biconditional in form, and may for example be only a one-way conditional" (1961, p. 355, fn.5). Both Richardson and Bickle quote this passage, but they neglect to mention Nagel's remark that immediately follows, namely: "But in this eventuality '*A*' is not replaceable by '*B*,' and hence the secondary science will not in general be deducible from a theory of the primary discipline" (ibid.). So Nagel's view seems to be that, for a general account of the scientific practice in question, one *cannot* have the reduction/deduction of a nonbasic theory by virtue of one-way conditionals because they do not guarantee that the nonbasic theory is replaceable by the basic theory.

Of course, Nagel scholarship aside, other positivists held the more liberal view in question (Hempel, 1966, p. 105). But, as the tradition developed, many others stipulated that the connecting principles must be biconditional laws in order to justify cross-theoretic identities. As Schaffner put it, "connectability later came to be best seen as representing a kind of 'synthetic identity'" (1993, p. 425). Indeed, one cannot achieve an important goal of reduction without such identities – a simplification in the world's ontology.[2]

Yet philosophers of science in the post-positivist era observed that, for many central cases of reduction, the terms of the basic reducing theory only *approximate* the terms of the reduced theory. Using an example from genetics, Schaffner remarked that, until the late 1950s, the gene was typically defined in three ways: "(1) the smallest segment of the chromosome that could undergo mutation, (2) the smallest segment of the chromosome that could recombine with its homologous chromosome in crossing over, and (3) the section of the chromosome functionally responsible for a unit of character" (1967, pp. 142–143). However, it was discovered that (1) through (3) have different physical referents, in particular, that (1) and (2) involve smaller sequences of DNA. Accordingly, geneticists in the 1960s began to speak of "mutons," "recons," and "cistrons," respectively, where only the cistrons play roughly the role of Mendelian genes. Indeed, the notion of a gene has been further revised to include coding in RNA (see Weber, 2005)

These developments were of some consequence. For it was then apparent that what can be derived from molecular genetics is a *revised* or *corrected* genetic theory and not the original theory developed by Mendel. Thus, Schaffner proposed an account of "approximate reduction," supplying a set of formal conditions for the derivation of a corrected theory that is strongly analogous to the original reduced theory by means of connecting principles which preserve referential identity (1967, p. 144). Along with the ideally exact form of reduction, Schaffner's approximate reduction can be conveniently summarized as follows:

> S DEDUCTION: either the original reduced T_R or a strongly analogous corrected T_R^* is deduced from the basic reducing T_B.

Moreover, because some scientists use the terminology of reduction to cover cases where the ontology of the original target theory is retained as well as cases where the ontology of the original target theory is replaced, Schaffner later extended his account to incorporate theory replacement. Specifically, in addition to either the direct (for T_R) or approximating (for T_R^*) derivability conditions of *S Deduction*, Schaffner proposed that theory replacement involves the derivation of an "experimental arena" for the original target of reduction T_R – a "domain" in Dudley Shapere's (1974) sense – which is the set of experimental results associated with T_R that are better accounted for by the basic theory (Schaffner, 1977, pp. 148–151; also 1993, pp. 427–432). Viewing T_R^* in terms of a lesser set of experimental results means that, for theory replacement, the deductive consequence of the basic theory is still connected with but no longer strongly analogous to the original T_R. The result is a "continuum of reduction relations" represented by a "General Reduction-Replacement Model" (Schaffner, 1977, pp. 148, 149). It can be captured by the following convenient summary of his formal requirements for both approximate reduction and replacement:

> S CONTINUUM: there is a continuum of strong to weak analogies between the reduced T_R and the corrected T_R^*, with the strong relations justifying retention and the weak relations justifying replacement of the ontology of T_R.

3. New Wave Permutations on the Schaffner Theme

Much discussion about reduction in contemporary philosophy of science has concerned "New Wave" theories developed and defended by Paul and Patricia Churchland, Clifford Hooker, and John Bickle.[3] These accounts are based upon the framework of Schaffner's General Reduction-Replacement model, but they include a number of distinctive features. More specifically, the Churchlands, Hooker, and Bickle accept the framework of a derivational model, the role of a corrected theory in deduction, and the continuum of cases from reduction to replacement. But they also believe that Schaffner's account is too permissive by allowing the corrected theory to contain elements of the uncorrected target for reduction. Hence, advocates in the New Wave place a condition on reduction that is central to their approach, namely, *a base-level constraint on the resources used in the construction or development of the corrected image T_R^**. Paul Churchland says that:

> [A] successful reduction ideally has the outcome that, under the term mapping effected by the correspondence rules, the central principles of T_R (those of semantic and systematic importance) are mapped onto general sentences of T_B that are theorems of T_B. Call the set of such sentences T_R^*. This set is the image of T_R within T_B . . . on the account given above it is not the reduced theory, T_R, that is deduced from the principles of T_B, as some other accounts have it. What is deduced from T_B is rather the set T_R^*, an equipotent image of T_R within the idiom of T_B. (P. M. Churchland, 1979, pp. 81, 83, with a change in the subscripts)

Churchland stipulates that the corrected image T_R^* must be part of the basic theory, being "general sentences of T_B" that constitute an "equipotent image of T_R within the idiom of T_B." Or, as Hooker describes it:

> Within T_B construct an analog, T_R^*, of T_R under certain conditions C_R such that T_B and C_R entails T_R^* and argue that the analog relation, AR, between T_R and T_R^* warrants claiming (some kind of) reduction relation R, between T_R and T_B. (1981, p. 49)

And again, regarding the constraint in question, Bickle labels the corrected image "I_B" to underscore its base-level nature, and he emphasizes the contrast with Schaffner's view:

> It is important not to confuse Hooker's deduced image I_B with Schaffner's corrected version of the reduced theory T_R^*. Hooker's I_B is characterized completely within the framework and vocabulary of T_B; Schaffner's T_R^* is a corrected version of T_R, and so is characterized (at least in part) out of the resources and vocabulary of the reduced theory. (2003, p. 17)

More precisely, as I read Schaffner, his account does not *require* that T_R^* be characterized (even in part) out of the resources and vocabulary of a nonbasic T_R, since it is framed in a general way to allow for "homogeneous" cases where the reduced and reducing theories share the same vocabulary (Schaffner, 1967, p. 144). Even so, Schaffner's account *allows* that T_R^* be characterized out of the resources and vocabulary of a nonbasic and heterogeneously specified T_R. But the New Wave constraint does not. It is uncompromising and exclusionary. It can be expressed thus:

> CH CONSTRUCTION: the language and concepts of the basic reducing T_B, not the original reduced T_R, must supply the resources for constructing the corrected image T_R^*.

CH Construction is the main difference between Schaffner's General Reduction-Replacement model and the basic New Wave approach.[4] I have previously argued that this difference should be discounted, since (a) some of *CH Construction's* advertised virtues are shared by competing theories, (b) the exclusionary demands of *CH Construction* appear to conflict with the co-evolution of theories where terms in the corrected theory T_R^* have developed out of the conceptual resources of both parent theories T_B and T_R, and (c) the adoption of *CH Construction* is unduly restrictive from a methodological point of view by ruling out reductive strategies that require aid from the concepts and vocabulary of the original reduced T_R (Endicott, 1998a, pp. 60–67).[5]

However, even if *CH Construction* is retained, I want to focus on the similarity between Schaffner's model and the New Wave account, a similarity that exists by virtue of the fact that advocates of the New Wave accept the *S Continuum*. As Paul Churchland puts it, because of the differing degrees of similarity between the original T_R and the corrected T_R^* from one scientific case to another, "we must be prepared to count reducibility as a matter of degree. Like translation, which may be faithful or lame, reduction may be smooth, or bumpy, or anywhere in between" (P. M. Churchland, 1979, p. 84). Or, as Patricia Churchland says:

> The evolving unifications seen in science therefore encompass not only smooth reductions with cross-theoretic identifications but also rather "bumpy" reductions where cross-theoretic identifications are problematic and involve revision of the old theory's concepts, and outright elimination with no cross-theoretic identifications at all. (P. S. Churchland, 1986, p. 284)

Likewise, because of the variable analog relation between T_R and T_R^*, Bickle says that "there is a spectrum or continuum of cases lying between the smooth and bumpy endpoints" from "perfect retention" to "total replacement" (1998, p. 30).

4. An Exhaustive Reading of Reduction or Replacement

Both Schaffner and the New Wave appear to make an assumption that initially seems appropriate for the philosophical task at hand, namely, they appear to assume that *a comprehensive account of scientific unification will only describe nonbasic theories that are either wholly reduced or replaced, or broken down into parts that are either reduced or replaced.* Put in a different way, there is no place for irreducible and irreplaceable theories or irreducible and irreplaceable parts of theories on the *S Continuum*. The assumption seems clear in Schaffner's case, since the formal conditions of his General Reduction-Replacement model are a simple disjunction of the conditions for approximate reduction and the conditions for theory replacement (Schaffner, 1977, p. 149; 1993, p. 429). Of course Schaffner permits "partial reduction" in a sense that will "allow for the possibility of a partially adequate *component* of T_R being maintained" (1977, p. 148, with a change in the subscript). But the implication seems to be that the adequate parts are maintained through reduction while the inadequate parts are naturally replaced – no adequate parts that are unreducible and irreplaceable on the *S Continuum*.

Or consider how the New Wave appears to treat the continuum of cases from reduction to replacement. On the one end there is perfect retention, and on the other end there is outright replacement, while the area *between* those endpoints is occupied by mixed cases of reduction and replacement. For example, Bickle emphasizes the middle ground of theory revision for a number of scientific cases. But *Bickle accepts no dualism* – be it a dualism of objects, properties, or laws (1998, pp. 6–14). Therefore, in cases of theory revision, the elements of the original nonbasic T_R are either retained through reduction or rejected, with a subsequent revision of concepts to reflect the distance between T_R and its corrected base-level counterpart T_R^* (ibid., p. 200). More precisely, according to Bickle, cases where T_R is closely analogous to its base-level counterpart call for homogeneous ORLs (ontological reductive links) that provide a relatively smooth reduction of the property elements in T_R, while middle cases where T_R is not smoothly reduced but not outright replaced call for mixed ORLs, where some property elements are reduced by homogeneous ORLs and the rest are treated in a way that respects "the eliminativist strand of revisionary physicalism" according to which the properties "are abandoned as lacking actual extension" (ibid., p. 202). Again, no irreducible and irreplaceable theories.

But why is this exhaustive reading of reduction or replacement problematic if the goal is to supply an account of *scientific unification* where, in fact, either a reduction or replacement is carried out?

5. Token Reductions as Partial Reductions

The assumption that a comprehensive account of scientific unification will only describe nonbasic theories that are either wholly reduced or replaced or broken down into parts that are either reduced or replaced is problematic even when the goal is to provide an account of scientific unification and even when unification is understood in terms of processes like reduction and replacement that simplify the world's ontology. Simply put, by giving place to type irreducible and irreplaceable theories, one is able to expand the range of partial reductions to include their *token reduction*. My working assumptions are:

(1) Reduction (versus replacement) is ontological unification *via* cross-theoretic identities.

(2) An adequate theory of reduction should provide an account that subsumes the widest range of cases involving ontological unification *via* cross-theoretic identities.

(3) Partial reductions that are accomplished through token identities provide ontological unification *via* cross-theoretic identities.

And hence:

(4) An adequate theory of reduction should provide an account that subsumes partial reductions that are accomplished through token identities.

To begin, there is to my knowledge no exhaustive study of partial reduction. But philosophers have used the term "partial reduction" in several senses. I will distinguish two. There is (A) a *Mereological* sense of "partial reduction" whereby not all the *parts* of a target theory are derived from a basic theory, and there is (B) a *Teleological* sense of "partial reduction" whereby not all the perceived *goals* for reduction have been met when a target theory is derived from a basic theory. Thus, according to the mereological notion, a theory is decomposable into parts, where only a proper subset of those parts is subject to reduction. This notion can be further subdivided into categories that subsume cases of retention or replacement with respect to the entire theory, depending upon whether a significant number of its derivable proper parts are central theoretical structures or merely the experimental arena better accounted for by its successor theory (cf. Sklar, 1967, p. 116, on the partial reduction of external sentences in a replaced theory).

In contrast, according to the teleological notion, there may be a partial reduction even when all the parts of a target theory are derivable from a basic theory. For example, Hartry Field (1972, p. 362) says that Tarski's semantic theory of truth provides only a "partial reduction" of truth to nonsemantic terms, since Tarski's theory utilizes primitive semantic notions like denotation for names and satisfaction for predicates. To understand why Field has the teleological notion in mind, let T_R represent the set of sentences that English speakers assert to be true, and let T_B represent Tarski's theory of truth for the English language. Field does not deny that all the parts of T_R can be derived from T_B in the sense that every sentence that English speakers assert to be true is a theorem of T_B. Rather, Field believes that, even though every one of those sentences is a theorem of T_B, the physicalist goal of providing a reduction of truth to nonsemantic or physical terms has not been met.

Here I am only interested in the mereological sense of "partial reduction." More specifically, I am only interested in the kind of partial reduction that occurs when the subset of T_R's derivable parts describe tokens of the types in T_R. Accordingly, T_R is type-irreducible but token-reducible. But T_R is not just type-irreducible. That is also true about replaced theories. Thus, my project is to incorporate their token reduction into a broad scheme of scientific unification from complete reduction to outright replacement, where the middle ground is *not* wholly occupied by nonbasic theories that are either reduced or replaced or broken down into parts that are either reduced or replaced, as it appears on the accounts developed by Schaffner and the New Wave.

Proposals about token reduction are not new. Fodor (1981a) outlined the general idea for the special sciences, and Steven Kimbrough (1979) developed and defended its application regarding the token reduction of genetics to molecular biology. Specifically, Kimbrough suggests a scheme of reduction functions for token cases. Changing his symbolism slightly, where "F" is a predicate of a nonbasic type-irreducible but token-reducible theory T_R, "P" is a predicate of a base-level token-reducing theory T_B, "a" is an individual constant in T_R, and "b" is an individual constant in T_B, the reduction functions are:

1. (for some x, y) [(Fx and Px) and (x = y)],
2. a = some x such that Fx,
3. b = some x such that Px,
4. a = b. (Kimbrough, 1979, pp. 403–404)

6. Schaffner's View of Token Reduction

Curiously, Schaffner acknowledges token reduction and explicitly cites both Fodor and Kimbrough, but he does not attempt to extend the range of partial reduction by such means. Instead, he says "there is some truth in Fodor's speculations and in Kimbrough's contentions," but he believes that "they are overstated in connection with reduction in the biomedical sciences generally and particularly in the area of genetics" (1993, p. 463). Yet his reservations appear to miss the mark. On the one hand, and perhaps because he is accustomed to thinking in terms of the traditional derivation of types, Schaffner says:

> First, it should be noted that, in genetics, it is not specific individuals demarcated in space and time that are identified; rather, the *lac p* and *o* genes present in many individual bacteria are identified with DNA sequences presumed to be repeated in many (a potentially infinite number of) instances. (ibid.)

But that surely *is* an overstatement, since (a) a complete formalization of genetic science contains singular terms, bound variables, and constants for individuals, and since (b) the broader practice of genetic science must look at specific individuals and their specific genetic materials, otherwise the theory would be useless. At best, Schaffner has only observed that tokens are not types.

On the other hand, Schaffner takes a different line by reinterpreting the Fodor–Kimbrough token reduction in terms of a more restricted range of (type) generalizations for the kind of middle-sized objects befitting the biological sciences. He goes on to say:

> [T]here is an element of truth in Kimbrough's concerns. Reduction in genetics and, I suspect, much of biology and medicine appears to be more specific; it does not yield the broad generalizations and identities one finds in physics. This, I think, is due in part to the systems one studies in biology, and also to the kinds of theories one finds at what I have termed [in a previous chapter] the "middle range." (1993, p. 465)

And again, in his summary remarks: "a less misleading gloss of 'token-token' in theses contexts would be to construe the term 'token' as a 'restricted type' in accordance with the view of theories developed in [previous] chapters" (ibid., p. 466). But, as Schaffner explains them, the restricted types in question are simply the types that are not as universal in their application as the types of basic physics, specifically, the types found in biology and the other special sciences (ibid., p. 97). So Schaffner comes full circle to the types of biology and the special sciences, and he nowhere explains how *their* tokens can be construed as restricted types, which seemed to be the proposal at issue.

Of course, others like David Lewis (1969) and Jaegwon Kim (1993b) postulate more restricted special science types to achieve the purpose of reduction. More accurately put, they postulate species-specific and even individual-time-specific terms in the special sciences that nonrigidly denote physical properties. So, if this narrow reductive strategy is successful, there is no need to speak of a mere token reduction of irreducible types. Rather, there is a complete reduction of types, construed in the species or individual-and-time restricted way. But others have argued, I think successfully, that species-specific terms fail to accommodate the inter-theoretic cross-classification that results from multiple realizability within individuals (see Horgan,

2001). Moreover, I have argued that the more restricted individual-time-specific terms are virtually indistinguishable from singular terms that pick out token events, and hence they fail to meet an essential desiderata for kind terms in the sciences, namely, being generalizable in the way needed for scientific explanation and prediction (Endicott, 1993, p. 317).

So Schaffner does not explicitly utilize a scheme of token reduction for type-irreducible theories. However, his philosophical attitude about type-irreducible theories permits such a scheme. For Schaffner is a methodological pluralist who accepts higher-level presently irreducible theories on pragmatic grounds but denies their irreducible status "in principle." In his words, Schaffner favors a "pragmatic, holistic, but in principle reductionist approach" that construes "theories in biology and medicine as essentially (for the present and foreseeable future) interlevel" (1993, pp. 413–414). Indeed, Schaffner's more recent work reflects a "reappreciation of the complexity by geneticists in the 1990s" that antireductionists had previously stressed about "many–many" mappings between genetics and molecular theory (2002, p. 323).[6] So his "in principle" reductions are far removed from present scientific reality which contains type-irreducible theories, and that is why my proposal to incorporate their token reduction, when developed, will constitute a more or less friendly amendment to Schaffner's General Reduction-Replacement model of scientific unification. Keep type-irreducible theories for pragmatic purposes, and interpret the subsequent token reduction functions accordingly – realistically where the world complies and instrumentally otherwise.

Why did Schaffner not explicitly incorporate a scheme of token reduction into his General Reduction-Replacement model if his methodological pluralism allows it? Perhaps Schaffner thought that the token reduction of type-irreducible theories is trivial. As William Wimsatt put the point with respect to psychological theory:

> Without type-correspondence, property identification seems to be ruled out, and about the only kind of identity left is "stuff" identity – roughly, that the stuff with the psychological properties is the same stuff as the stuff with the physical properties. Philosophers, concentrating on ontological dividends, have found this to be uninteresting and trivial. It seems like a common-sense conclusion that we hardly need scientific sophistication or data to embrace. (1976, p. 225)

But one would be wise not to accept Wimsatt's casual record of common sense, since the token reduction of type-irreducible theories is hardly trivial. Without type-correspondence, there is no guarantee that two theories drawn from different scientific levels of inquiry will carve up the "stuff" of the world into the same naturally isolated particulars that are subject to the same explanations and predictions. To use a well-worn example, many tokens of economic theory – monetary transactions, financial institutions, aggregate supplies, the economic cost for Enron from 1997 to 2000, the US labor force in 2006 – are not naturally isolated particulars from the vantage point of neuroscience, chemistry, or basic physics. I will return to this problem in the final section. But, in my estimation, the reason Schaffner did not incorporate the token reduction of type-irreducible theories into his model of scientific unification is that it is built upon an assumption that excludes viable type-irreducible theories. Specifically, Schaffner's model is built upon the assumption that the corrected theory "bears close similarity" to the original target for reduction (Schaffner, 1967, p. 144), which, when expanded into a continuum that includes theory replacement, implies the following *Similarity Assumption*:

> S ASSUMPTION: similarity is the parameter which determines the place of T_R along the *S Continuum*, with strong to weak similarity relations between T_R and its base-level derivable counterpart T_R^* determining reduction to replacement, respectively.

But note the consequence. The more *dissimilar* T_R is with respect to T_R^*, *the more appropriate it is for replacement*. Yet, by virtue of the complex, many–many mappings between the kind terms of special science and the kind terms of physical science, *type-irreducible special science theories are exceedingly dissimilar to anything derivable from a base-level physical theory*. Hence, by the *Similarity Assumption*, these irreducible theories must be replaced – surely the wrong result. Happily, in two sections hence I will propose a broader continuum from reduction to replacement where "similarity" is replaced by formal conditions of "connectedness" to base-level theories, which allows viable or at least pragmatically useful type-irreducible theories to be retained in spite of their dissimilarity to base-level theories.

7. The New Wave View of Token Reduction

At first glance, some members of the New Wave appear to look favorably upon the practice of token reduction. In particular, Hooker (1981, pp. 504–507) outlines an account of "function-to-structure token reduction," and Bickle (1998, pp. 155–163) endorses it. But this New Wave token reduction differs from the standard variety proposed by Fodor and Kimbrough. Simply put, unlike the standard account, New Wave token reduction appears to be a guised form of eliminativism. I have presented the details elsewhere, so I will be brief (Endicott, 2001, pp. 388–391; see also Wright, 2000). Hooker illustrates New Wave token reduction by employing something roughly parallel to the familiar tripartite taxonomy employed within cognitive science – the semantic (L_1), syntactic (L_2), and the physical mechanistic (L_3) levels of description. He constructs a target theory T out of higher-level L_1 and L_2 predicates, and the basic-level theory T* out of mechanistic L_3 predicates, claiming that:

> Systems of a type S of class T are contingently token/token identical with systems of type S' in class T* $=_{df}$ every instance (token) of a type S system externally classified as in class T is contingently identical with some instance (token) of a type S' system externally classified as in class T*. (1981, 504)

So far so familiar. But the standard idea of token identity involves *the same object exemplifying two distinct properties*, a token identity with types distinct. In an inter-level case, this means *property dualism*, the same object having both an irreducible higher level property and a lower-level property. Yet the New Wave rejects property dualism.

So New Wave token reduction must be something else called "token reduction," and it seems to be a version of eliminativism. When considering the possibility that each functional property is either (i) type-reducible by being directly identified with a mechanistic property or (ii) token-reducible by having its instance identified with an instance of a mechanistic property, Hooker says that "there are no properties corresponding to predicates falling under case (ii) above if by this one means a single property common to all instances," and that such cases may require "resisting putative $L_1 + L_2$ semantics"

(1981, p. 507). Similarly, Bickle says: "In such cases we conclude that the functional-level predicates fail to denote" (1998, p. 162). So this is no familiar token reduction where the same object exemplifies two distinct properties. Only one property is retained, and thus only its tokens remain.

The latter point is worth emphasizing. On the New Wave scheme *there are no tokens of the type-irreducible properties.* As I stated before, if there is no phlogiston, then nothing is token identical with phlogiston (Endicott, 2001, p. 390). Aside from the types and tokens of T's observational consequences that are better explained by T*, there are just the items picked out by mechanistic-level predicates, including perhaps some functional-level predicates that have been semantically "reconstructed" to denote mechanistic-level properties (Hooker, 1981, pp. 507–512). So the label "token reduction" is justified, apparently, by the lingering use of some reconstructed functional predicates.

This eliminative interpretation of New Wave token reduction also accords well with the New Wave tendency to view higher-level irreducible theories as radically false targets for elimination.[7] For example, Churchland (1981) has famously argued that propositional attitude psychology should be eliminated. And, although Bickle (1998, pp. 205–206) describes a more nuanced revisionary position according to which folk psychology errs about the fine-grained causal structure but succeeds in describing the "gross abstract structure" of cognition, given the exhaustive reading of "reduction or replacement" discussed earlier, Bickle must believe that the limited set of true and successful descriptions from folk psychology pick out properties that reduce to neurochemical properties.

However, like Schaffner's pragmatic attitude toward presently type-irreducible theories, the outlook of the New Wave toward the broader practice of science also permits the use of type-irreducible theories for token reduction. Thus, Bickle's seemingly ruthless claim that psychological explanations "become otiose" and are rendered "impotent" by neuroscientific theories presupposes an "*accomplished* (and not just an anticipated) cellular/molecular explanation" (2003, p. 110, original emphasis). Moreover, even when base-level (presumably reductive) explanations have been achieved, Bickle allows "some residual, purely heuristic tasks" (1998, pp. 205–206). Consequently, it is consistent with Bickle's views that when base-level reductive explanations have *not* been achieved, one may retain presently type-irreducible theories for heuristic reasons. But this kind of philosophical outlook is no eliminativism. For example, instrumentally interpreted theories are neither reduced nor replaced. Like their realist counterparts – property dualist theories – they are retained in their presently irreducible forms precisely because of their convenience and overall utility. My proposal will thus make room for more conciliatory pragmatic and instrumental attitudes by expanding the *S Continuum* so that it is no longer exhausted by theories that are either wholly reduced or replaced or broken down into parts that are reduced or replaced.

8. The Space between Reduction and Replacement

To avoid confusion with the kind of "retention" that occurs with type reduction, let "reception" designate the space between reduction and replacement where type-irreducible theories are retained for the purpose of token reduction. Moreover, call the continuum that contains a space for the reception of type-irreducible theories the *Continuum from Reduction to Reception to Replacement.* I will sketch, in programmatic fashion, some necessary conditions that a pair of theories must satisfy for placement along this *3R Continuum.* Intuitively speaking, the *3R Continuum* registers the degree of "connectedness" between basic and nonbasic theories. Being mindful that the meaning of the subscript "$_R$" for theory T now varies in one of three ways – reduced, received, and replaced – the conditions for placement are as follows.

R1. Nonbasic T_R is a reduced theory. The general terms of a nonbasic theory positioned at the reductive end of the *3R Continuum* are connected to the general terms of the base-level theory in the way required by *S Deduction.* There is a direct (for T_R) or approximating (for T_R^*) derivation from a base-level T_B by means of connecting principles that express property identities between the two theoretical domains. Since type identities are established, the token identities are secure.

R2. Nonbasic T_R is a received theory. Approximating revisions no longer secure type reduction. The general terms of a nonbasic theory positioned at the receptive area of the *3R Continuum* are no longer connected to the general terms of the base-level theory by principles that express property identities between the two theoretical domains. But there exist connections between the properties supplied by nonreductive determinative relations like supervenience (Kim, 1993a; Horgan, 1993) and realization (Endicott, 2005). So the properties of T_R, or the properties of T_R^* when the former has been corrected, are determined by the properties of the base-level T_B in a way compatible with their nonidentity. Yet there remain connections between the singular terms of the nonbasic theory and the singular terms of the basic theory that establish token identities, and these connections conform to the token reduction functions presented earlier.

R3. Nonbasic T_R is a replaced theory. Whether general or singular, very few terms of the nonbasic theory positioned at the replacement end of the *3R Continuum* are connected to the terms of the base-level theory. The ontology is no longer anchored into the world. The types and tokens of T_R are thus eliminated in favor of the types and tokens of the base-level T_B that, by virtue of its superior resources, provides a better explanation of the observable phenomena originally targeted by the displaced T_R.

So I have replaced talk of strong to weak analogies between nonbasic and base-level derivable theories with a more precise analysis in terms of the kind of connections between them, and I have replaced the exhaustive reading of reduction or replacement with a broader set of conditions that allow the reception of type-irreducible theories that are subject to token reduction. However, my aim is *not* to portray a broad nonreductive view of the sciences but to expand the range of partial reductions *via* token identities. Put in a different way, I have not simply reinterpreted "unification" in a nonreductive way that allows for the supervenience or realization of type-irreducible theories. That misconceives the point entirely. As a conservative extension of the models under consideration, "unification" still means ontological unification *via* cross-theoretic identities. Thus, in the service of that august reductive goal, and in summary fashion:

> 3R CONTINUUM: there is a continuum of connections between the nonbasic T_R and its base-level derivable counterpart T_R^*, with type-identities justifying type reduction, token identities justifying token reduction, and the lack of such identities justifying replacement of the ontology of T_R.

The *3R Continuum* describes a very broad inter-level slice of the world. But this is as it should be. Reduction and other forms of scientific unification cannot be considered in isolation from the larger body of scientific practice in which they occur.

9. Concluding Remarks

My primary goal has not been to place particular theory pairs on the *3R Continuum*, or contest their placement. Granted, particular cases and general theories must be brought into reflective equilibrium. But my focus has been on developing a general theory of scientific unification from reduction to replacement by exploring how the possibilities of token reduction can be situated within that framework. Yet the matter of token reduction is not trivial, and there are a number of problems that should be resolved once the general outline of a model has been presented. Thus, I will end by citing one problem, then briefly indicate how it might be addressed, leaving a fully adequate discussion of these points for another time.

Simply put, there is a problem about *token correspondence for inter-level theories.* As previously mentioned, without type-correspondence, there is no guarantee that two theories drawn from different scientific levels of inquiry will carve up the world into the same naturally isolated particulars that are subject to the same explanations and predictions. Focusing on the psychophysical case, Stephen Stich presents a list of authors from Winograd to Minsky who offer models of cognitive phenomena wherein "no single component or naturally isolated part can be said to underlie the expression of a belief or desire" (1983, p. 240).[8] Indeed, semantic networks are often thought to display the phenomenon in question, since there may be large areas of computational activity that are shared by several belief states all at once. So, if particular beliefs exist, they arise from the base-level states in a more holistic fashion. Moreover, a general token correspondence problem seems to be a natural consequence of the fact that, given a mereological account of levelhood from the microphysical to the macrosocial, each level *n* of scientific inquiry typically involves much larger aggregates of particulars than the more basic level *n – 1*.

In response, one might endorse a solution that utilizes constructivist procedures, composing aggregates of base-level objects in a way that preserves token identities. The smaller particulars of the base-level theory are combined by logical operations into aggregates that are token identical with the particulars of the nonbasic theory. But, until more is said, these constructed aggregates may appear highly gerrymandered from the perspective of the base-level theory – like financial institutions and labor forces appear from the perspective of fundamental physics. I prefer a more scientifically inspired solution.

As a start, one might concede that token reductions are rarely accomplished because of the different goals and principles that operate at different levels of scientific inquiry. Perhaps token reductions apply only at the borderline of neighboring disciplines where interests converge. Nevertheless, there is a process that can bring them about, namely, the process of co-evolution when it adjusts two neighboring theories until a mechanism is discovered whereby some ontology of the one theory realizes or implements some ontology of the other theory.

More specifically, co-evolution is a process where the concepts and languages of two theories at different levels of inquiry develop together by their mutually influence upon each other (Wimsatt, 1976, pp. 230–237; see also McCauley, 1996). For example, neuroscientists employ psychofunctional criteria to guide the identification of brain structures, while cognitive scientists employ information about brain structures to guide their theories of information processing (there is a nice discussion in Bechtel & Mundale, 1999). This co-evolution may continue until, by the continued calibration of one theory to the other, a base-level mechanism is discovered which supplies the now naturally isolated unit for token reduction. That is, the base-level mechanism yields a token of the pertinent nonbasic realized type. This idea can be captured by a familiar form of functional explanation. Employing the symbolism used for the token-reduction functions presented earlier, one begins with a *functional analysis*: x has a nonbasic property F = x has something that plays a given functional role R. Then one utilizes *the empirical discovery* that a type of base-level physical mechanism P plays the functional role R. And from this one can draw *the conclusion*: if x has the mechanism property P then x has the nonbasic property F. In other words, one can deduce that the *same x* has properties F and P – the desired token identity. To illustrate with a concrete case, consider the Crick–Koch Hypothesis (1990) that the mechanism for human visual awareness is 40–70 Hz neural oscillations in the human cortex:

(i) x has visual awareness = x has something that plays the functional role of controlling information from the visual input system in a global fashion with respect to attention, short-term memory, and behavioral outputs.
(ii) x has 40–70 Hz oscillations of neurons in the human visual cortex ⇒ x has something that plays the functional role of controlling information from the visual input system in a global fashion with respect to attention, short-term memory, and behavioral outputs.
(iii) x has 40–70 Hz oscillations of neurons in the human visual cortex ⇒ x has visual awareness.[9]

Again, *the same x* has the neural property and the functional awareness property. Interestingly, the foregoing argument also reveals a merger of the two traditions of reduction mentioned at the outset of this chapter. Reduction as mechanistic explanation and reduction as ontological unification *via* cross-theoretic identity meet at the point of token reduction.

Acknowledgment

I would like to thank David Austin and the editors of the present volume for their helpful comments on an earlier draft of this chapter.

Notes

1 Sklar gives a more cautious reading, saying that Nagel's text "appears" to signal a change: "Originally Nagel insisted upon theses correlatory hypotheses being universally quantified bi-conditionals, with one side of the bi-conditional containing as its only descriptive term one of the terms peculiar to the reduced theory. Subsequently he appears to have weakened this condition, allowing the laws to take other forms as well" (Sklar 1967, p. 118).
2 Patricia Kitcher (1980) also emphasizes this point in her response to Richardson. I add that both Schaffner's General Reduction-Replacement model and the basic New Wave approach likewise associate reduction with cross-theoretic identities. For Schaffner, the identities surface in the interpretation of the connecting principles that facilitate the derivation of either the original or corrected targets for reduction (1967, p. 144). For the New Wave, the identities surface as a consequence of a com-

parative smooth analogy between the original and corrected theories (P. M. Churchland, 1985, p. 11; Hooker, 1981, p. 45; P. S. Churchland, 1986, p. 284; Bickle, 1998, pp. 77–78). Indeed, I have shown that these ontological consequences imply traditional biconditional bridge laws that license cross-theoretic property identities (Endicott, 1998a, pp. 67–72). Perhaps for this reason, Bickle (2003, pp. 31–39) has recently adopted a Carnapian-inspired, "internalist," "metascientific" attitude that rejects "metaphysical" identity questions altogether. I do not have the space to address Bickle's return to Positivist themes here.

3 The label was first applied by Bickle (1996, p. 57).

4 Other New Wave proposals are less central in the sense that they vary from one New Wave advocate to the next. For example, both Paul Churchland and Bickle adopt a nonsentential view of scientific theories. But Churchland (1990) prefers to think of them in the connectionist tradition as vectors through an abstract state-space, while Bickle (1998) prefers to think of them in the semantic tradition as model-theoretic structures.

5 Maurice Schouten has understandably requested that I clarify my charge about co-evolution, since advocates of the New Wave also accept the co-evolution of theories. My earlier argument was that their account of co-evolution appears inconsistent with *CH Construction*, given that the "mutual" interplay of co-evolution includes the top-down role of nonbasic concepts in constructing the corrected theory T^{*}_{R} (Hooker, 1981, pp. 513–514; Bickle, 1996, p. 76, 1998, pp. 148, 201). In the very least, if one accepts top-down influences of a nonbasic T_R on a resulting base-level T^{*}_{R}, then, as I stated before, *CH Construction*'s advertised exclusion of the concepts and language of T_R must appear superficial from the historical development of the sciences, since it ignores the interplay between basic and nonbasic theories (Endicott, 1998a, p. 66).

6 Hull (1974, esp. pp. 38–42) stressed the complicated many–many mappings for molecular genetics. For a general analysis of inter-level one–many and many–one relations, see Endicott (1994, 1998b).

7 Compared to the eliminative spirit that seems to underlie his earlier remarks on function-to-structure token reduction, Hooker's later work expresses a "naturalism" that does not require reduction, being consistent with "dualisms of many sorts" (1987, p. 261).

8 In addition to the sources cited by Stich, see also Horgan & Tye (1985). Given the failure of token correspondence, some have drawn antirealist conclusions about the ontology of the higher-level theory. But, as Horgan & Tye (1985, p. 438, fn.13) observe, that conclusion can be resisted.

9 Note that the second premise and the conclusion are couched in terms of one-way conditionals that express mere sufficient conditions, consonant with the type irreducibility of visual consciousness. If they were

biconditionals that justify the stronger connections of identity, one would have a type reduction of visual consciousness. Crick and Koch apparently take the mechanisms to license type reduction. I do not. I think there can be visual consciousness realized by some mechanism other than 40–70 Hz neural oscillations.

References

Bechtel, W., & Mundale, J. (1999). Multiple realizability revisited: Linking cognitive and neural states. *Philosophy of Science*, *66*, 175–207.

Bechtel, W., & Richardson, W. (1993). *Discovering Complexity: Decomposition and Localization as Strategies in Scientific Research*. Princeton: Princeton University Press.

Bickle, J. (1996). New wave psychophysical reduction and the methodological caveats. *Philosophy and Phenomenological Research*, *56*, 57–78.

Bickle, J. (1998). *Psychoneural Reduction: The New Wave*. Cambridge, MA: MIT Press.

Bickle, J. (2003). *Philosophy and Neuroscience: A Ruthlessly Reductive Account*. Dordrecht: Kluwer Academic Publishers.

Causey, R. (1972). Attribute-identities in microreduction. *Journal of Philosophy*, *69*, 407–422.

Churchland, P. M. (1979). *Scientific Realism and the Plasticity of Mind*. London: Cambridge University Press.

Churchland, P. M. (1981). Eliminative materialism and the propositional attitudes. *Journal of Philosophy*, *78*, 67–90.

Churchland, P. M. (1985). Reduction, qualia, and the direct introspection of brain states. *Journal of Philosophy*, *82*, 8–28.

Churchland, P. M. (1989). On the nature of explanation: A PDP approach. Reprinted in P. M. Churchland, *A Neurocomputational Perspective: The Nature of Mind and the Structure of Science* (pp. 197–230). Cambridge MA: MIT Press.

Churchland, P. M. (1990). On the nature of theories: A neurocomputational perspective. Reprinted in *A Neurocomputational Perspective: The Nature of Mind and the Structure of Science* (pp. 153–196). Cambridge MA: MIT Press.

Churchland, P. S. (1986). *Neurophilosophy: Toward a Unified Science of the Mind/Brain*. Cambridge, MA: MIT Press.

Crick, F., & Koch, C. (1990). Towards a neurobiological theory of consciousness. In N. Block, O. Flanagan, & G. Guzeldere (Eds.), *The Nature of Consciousness* (pp. 277–292). Cambridge, MA: MIT Press, 1997.

Dobzhanski, T. (1968). On some fundamental concepts of Darwinian biology. In T. Dobzhanski, M. Hecht, & W. Steere (Eds.), *Evolutionary Biology* (Vol. 2, pp. 1–34). New York: Appleton-Century-Crofts.

Endicott, R. (1993). Species-specific properties and more narrow reductive strategies. *Erkenntnis*, *38*, 303–321.

Endicott, R. (1994). Constructival plasticity. *Philosophical Studies*, *74*, 51–75.

Endicott, R. (1998a). Collapse of the New Wave. *Journal of Philosophy*, *95*, 53–72.

Endicott, R. (1998b). Many–many mappings and world structure. *American Philosophical Quarterly*, *35*, 267–280.

Endicott, R. (2001). Post-structuralist angst – critical notice: John Bickle, *Psychoneural Reduction: The New Wave*. *Philosophy of Science*, *68*, 377–393.

Endicott, R. (2005). Multiple realizability. In D. Borchert (Ed.), *The Encyclopedia of Philosophy* (Vol. 6, 2nd ed., pp. 427–432). Detroit: Macmillan Reference.

Field, H. (1972). Tarski's theory of truth. *Journal of Philosophy*, *69*, 347–375.

Fodor, J. (1981a). Special sciences (or: the disunity of science as a working hypothesis). Reprinted in J. Fodor, *Representations: Philosophical Essays on the Foundations of Cognitive Science* (pp. 127–145). Cambridge, MA: MIT Press. (Original work published 1974.)

Fodor, J. (1981b). Something on the state of the art. In J. Fodor, *Representations: Philosophical Essays on the Foundations of Cognitive Science* (pp. 1–31). Cambridge, MA: MIT Press. (Original work published 1974.)

Hempel, C. (1966). *Philosophy of Natural Science*. Englewood Cliffs, NJ: Prentice-Hall.

Hooker, C. (1981). Towards a general theory of reduction. Part I: Historical and scientific setting. Part II: Identity in reduction. Part III: Cross-categorical reduction. *Dialogue*, *20*, 38–59, 201–236, 496–529.

Hooker, C. (1987). *A Realistic Theory of Science*. Albany, NY: SUNY Press.

Horgan, T. (1993). From supervenience to superdupervenience: Meeting the demands of a material world. *Mind*, *102*, 555–586.

Horgan, T. (2001). Multiple reference, multiple realization, and the reduction of mind. In G. Preyer & F. Siebelt (Eds.), *Reality and Humean Supervenience: Essays on the Philosophy of David Lewis* (pp. 205–221). Lanham, MD: Rowman & Littlefield.

Horgan, T., & Tye, M. (1985). Against the token identity theory. In E. Lepore & B. McLaughlin (Eds.), *Actions and Events: Perspectives on the Philosophy of Donald Davidson* (pp. 427–443). Oxford: Basil Blackwell.

Hull, D. L. (1974). *Philosophy of Biological Science*. Englewood Cliffs, NJ: Prentice-Hall.

Kim, J. (1992a). Multiple realization and the metaphysics of reduction. In J. Kim, *Supervenience and Mind: Selected Philosophical Essays* (pp. 309–335). London: Cambridge University Press. (Original work published 1984.)

Kim, J. (1993b). Concepts of supervenience. In J. Kim, *Supervenience and Mind: Selected Philosophical Essays* (pp. 53–78). London: Cambridge University Press. (Original work published 1984.)

Kimbrough, S. O. (1979). On the reduction of genetics to molecular biology. *Philosophy of Science*, *46*, 389–406.

Kitcher, P. (1980). How to reduce a functional psychology? *Philosophy of Science*, *47*, 134–140.

Lewis, D. (1969). Review of *Art, Mind, and Religion*. *Journal of Philosophy*, *66*, 23–35.

Machamer, P., Darden, L., & Craver, C. (2000). Thinking about mechanisms. *Philosophy of Science*, *67*, 1–25.

McCauley, R. (1996). Explanatory pluralism and the co-evolution of theories in science. In R. McCauley (Ed.), *The Churchlands and Their Critics* (pp. 17–47). Oxford: Blackwell.

Nagel, E. (1961). *The Structure of Science: Problems in the Logic of Scientific Explanation*. New York: Harcourt, Brace, and World.

Richardson, R. C. (1979). Functionalism and reductionism. *Philosophy of Science*, *46*, 533–558.

Schaffner, K. (1967). Approaches to reduction. *Philosophy of Science*, *34*, 137–147.

Schaffner, K. (1977). Reduction, reductionism, values and progress in the biomedical sciences. In R. Colodny (Ed.), *Pittsburgh Series in the Philosophy of Science* (pp. 143–171). Pittsburgh, PA: University of Pittsburgh Press.

Schaffner, K. (1993). Reduction and reductionism in biology and medicine. In K. Schaffner, *Discovery and Explanation in Biology and Medicine* (pp. 411–516). Chicago: University of Chicago Press.

Schaffner, K. (2002). Reductionism, complexity and molecular medicine: Chips and the "globalization" of the genome. In M. Regenmortel & D. Hull (Eds.), *Promises and Limits of Reductionism in the Biomedical Sciences* (pp. 323–347). Chichester: John Wiley.

Shapere, D. (1974). Scientific theories and their domains. In F. Suppe (Ed.), *The Structure of Scientific Theories* (pp. 518–565). Urbana: University of Illinois Press.

Sklar, L. (1967). Types of inter-theoretic reduction. *British Journal for the Philosophy of Science*, *18*, 109–124.

Stich, S. (1983). *From Folk Psychology to Cognitive Science: The Case Against Belief*. Cambridge, MA: MIT Press.

Weber, M. (2005). *Philosophy of Experimental Biology*. Cambridge: Cambridge University Press.

Wimsatt, W. (1976). Reductionism, levels of organization, and the mind–body problem. In G. Globus, G. Maxwell, & I. Savodnik (Eds.), *Consciousness and the Brain: A Scientific and Philosophical Inquiry* (pp. 196–267). New York: Plenum.

Wright, C. (2000). Eliminativist undercurrents in the new wave model of psychoneural reduction. *Journal of Mind and Behavior*, *21*, 413–436.

8

REDUCING PSYCHOLOGY WHILE MAINTAINING ITS AUTONOMY VIA MECHANISTIC EXPLANATIONS

William Bechtel

1. Introduction

Two related legacies of mid-twentieth-century philosophy of science continue to be entrenched in the philosophy of the cognitive sciences – the deductive-nomological model of explanation and the account of theory reduction. Philosophers talk as if the way to explain mental activities is by subsuming them under laws. Although psychologists sometimes advert to laws (as Cummins notes, usually designating them *effects*), these are seldom appealed to in order to explain their instances but rather to identify the phenomena (the regularities) requiring explanation (Cummins, 2000; Bechtel & Abrahamsen, in press). When they offer explanations, psychologists like biologists propose accounts of the mechanism responsible for the phenomenon. I will say more about mechanistic explanations below, but first I introduce the second legacy of mid-twentieth-century philosophy of science, the model of theory reduction. On this traditional philosophical account of interlevel relations in science (Oppenheim & Putnam, 1958; Nagel, 1961) the laws of a higher-level science (e.g., psychology) are reduced by being derived from the laws of the lower-level science (e.g., neuroscience) together with bridge principles and boundary conditions (for a discussion of the traditional theory reduction model, variants on it, and criticisms of it, see Bechtel & Hamilton, in press; McCauley, in press). According to this view, reduction is antithetical to any claims for autonomy of higher levels of organization and the sciences that investigate them. In order to derive one set of statements from another, the former cannot assert anything not already asserted by the latter and hence is redundant to it. Hence, it makes no autonomous contribution.

Given the theory reduction framework, theorists defending the autonomy of psychology and other higher-level sciences have argued against the possibility of reduction. They have targeted their criticisms on the possibility of bridge principles linking the vocabulary of the lower- and higher-level sciences, maintaining that higher-level phenomena are multiply realized by an extremely broad and not well delineated set of lower-level realizers (Fodor, 1974; Putnam, 1967). For over twenty years philosophers took multiple realizability as an obvious truth, but recently a number of philosophers have called into question its significance for science (Bechtel & Mundale, 1999; Bickle, 2003; Polger, 2004; Shapiro, 2004). Putnam appealed to examples such as hunger and pain, psychological phenomena which he claimed were realized in a huge diversity of animals despite radical differences in their brains. But it is important to note that hunger and pain behaviors vary radically across species, so if we view psychological processes at even a moderately fine grain, the psychological processes that are realized in different species are only similar, not identical. On the other hand, neuroscientists operate comparatively in identifying brain processes, treating as comparable brain activity in different species. If one uses a fine-grained account of both mental and neural processes, there is no evidence of the same mental state being realized in different ways. If, on the other hand, one adopts a coarse-grained account of both mental and neural processes, then, given the highly conserved nature of biological mechanisms, the different realizations of higher-level phenomena will themselves tend to be grouped in a common type. Accordingly, on neither a fine- or a coarse-grained account do appeals to multiple realizability provide a compelling argument against reduction and for the autonomy of higher-level explanations.

The theory reduction model, however, is much stronger than what scientists generally have in mind when they speak of reduction. For many scientists, research is reductionistic if it appeals to lower-level components of a system to explain why it behaves as it does under specified conditions. This sense of reduction is captured in the accounts of mechanistic explanation presented in the next section. I will argue in subsequent sections that the reductions achieved through mechanistic explanations are in fact compatible with a robust sense of autonomy for psychology and other special sciences, albeit a sense of autonomy no reductionist except one seeking hegemony for the lower level (Bickle, 1998, 2003) should have any desire to deny. This autonomy maintains that psychology and other special sciences study phenomena that are outside the scope of more basic sciences but which determine the conditions under which lower-level components interact. In contrast, the lower-level inquiries focus on how the components of mechanisms operate when in those conditions. Importantly, this defense of autonomy does not require appeal to multiple realizability, but only to the fact that investigations at higher levels of organization provide information additional to that provided by the account of how the parts of a mechanism operate.

Just as this notion of reduction is much weaker than that offered by the theory reduction model, so also is the notion of autonomy less than that defended by the model's opponents. In making room for the autonomy of psychology or other special sciences, I am not arguing that these inquiries should be pursued in ignorance of lower- (or higher-) level inquiries. Sometimes knowledge about the components of a mechanism can guide inquiry into how the mechanism engages its environment and when such knowledge is available, ignoring it is foolhardy. The same, though, applies in the opposite direction – knowing how a mechanism behaves under different conditions can guide the attempt to understand its internal operation. Inquiries at different levels complement each other both in the sense of providing information that cannot be procured at other levels and also in the sense of providing information that can limit the range of possibilities at other levels.

2. Mechanisms and Mechanistic Explanations

Within the life sciences, explanation frequently takes the form of identifying the mechanism responsible for a phenomenon of interest

– circulation of the blood, photosynthesis, protein synthesis, reproduction, etc. Although there are differences among the various accounts of mechanism and mechanistic explanation that have been advanced (Bechtel & Richardson, 1993; Bechtel & Abrahamsen, 2005; Glennan, 1996; 2002; Machamer, Darden, & Craver, 2000), these are inconsequential for the question of the relation of reduction and autonomy. My preferred characterization of a mechanism is that

> a mechanism is a structure performing a function in virtue of its component parts, component operations, and their organization. The orchestrated functioning of the mechanism is responsible for one or more phenomena. (Bechtel & Abrahamsen, 2005)

Explanation then consists in representing (sometimes verbally, but often in diagrams or in a computational model) the mechanism responsible for the activity and showing how it accounts for the phenomenon.

As in biology, most explanations in psychology involve the identification and characterization of a mechanism responsible for the phenomenon of interest – decision making, memory encoding and retrieval, language comprehension and production, etc. (Wright & Bechtel, in press). Until recently, psychologists have lacked the resources to identify the parts of the brain responsible for the various operations invoked in a mechanistic account, and have settled for identifying operations and modeling their interactions in generating the phenomenon. The information processing models they advanced proposed sequences of operations on informational structures (representations) that would account for the phenomenon of interest (e.g., problem solving). These models were generally tested using behavioral measures such as reaction times (Posner, 1978). With the advent of cognitive neuroscience, mechanistic explanations of mental phenomena have increasingly included identification of the brain parts responsible for the component operations.[1] Techniques such as neuroimaging enable researchers to identify the brain regions involved in executing a cognitive task. The goal of such research is not just to learn where operations occur, but to use such knowledge to further constrain and revise proposed accounts of mechanisms (e.g., by discovering that what were taken to be two entirely separate cognitive activities invoke the same neural process and asking how the same cognitive operation figures in both activities).

To give some substance to the idea of mechanistic explanation, I will develop a few central features of how mechanistic explanations commonly develop. First, a researcher begins by delineating the phenomenon of interest. One product of the cognitive revolution of the 1950s and 1960s was the differentiation of types of memory. Miller, Norman, and Sperling distinguished echoic, short-term, and long-term memory in terms of their behavioral characteristics (Neisser, 1967), while Tulving and his collaborators differentiated various forms of long-term memory (semantic, episodic, and procedural) (Tulving, 1983). In addition, investigators distinguished different time stages in memory – encoding, storage, and retrieval. Although some investigators were interested in characterizing the mechanisms involved (Atkinson & Shiffrin, 1968), this inquiry focused on differentiating and characterizing various memory phenomena. The fact that the ability to encode new episodic memories was selectively destroyed in Scoville's patient HM both secured the delineation of episodic memory encoding as a distinct phenomenon and implicated the hippocampus in it (Scoville & Milner, 1957).

All of these steps, however, are preliminary to advancing mechanistic explanation. This requires decomposing the mechanism into component parts and operations and localizing each operation in the appropriate part. Although psychologists have developed tools to demonstrate when different tasks involve different operations (Kolers & Roediger, 1984; Roediger, Buckner, & McDermott, 1999), they have made far less progress in specifying what the operations are (Bechtel, 2005). In some cases the determination of the relevant brain structure can provide clues as to the nature of the operations. For example, the determination that the hippocampus plays a role in the encoding of long-term memories inspired researchers to investigate whether the neuroarchitecture of the hippocampus might provide clues to how it realized this phenomenon. The hippocampus is comprised several different regions, each of which has a distinctive architecture (the relations between these areas are shown schematically in Figure 8.1). The dentate gyrus, for example, has ten times the number of neurons as the entorhinal cortex from which it receives inputs. This, together with the fact that only a few cells in the dentate gyrus respond to a given input stimulus, suggested that the dentate gyrus might serve to maintain separation between similar inputs and facilitate remembering events as distinct. The dentate gyrus projects

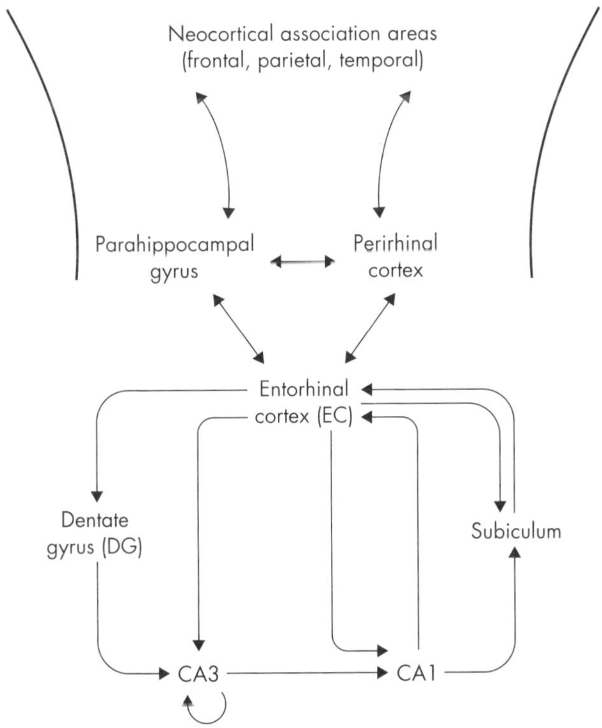

Figure 8.1 Schematic diagram of the hippocampal system. Information from widespread areas of neocortex converge on the parahippocampal region (parahippocampal gyrus, perirhinal cortex, and entorhinal cortex, EC) to be funneled into the processing loops of the hippocampal formation. The tightest loop runs from EC into the core areas of the hippocampus (CA1 and CA3) and back. An alternative route to CA3 goes through the dentate gyrus and an alternative route back to EC from CA1 goes through the subiculum, which is not part of the hippocampus proper. Not shown are a number of subcortical inputs and details of pathways and their synapses.

to the CA3 fields, whose neurons are highly connected to each other via recurrent loops, which suggests that they may compute similarity between patterns (Redish, 1999).

Beyond distinguishing phenomena and decomposing the responsible system functionally and structurally, mechanistic explanation requires researchers to determine how the various component parts are organized such that the operations are coordinated appropriately to realize the overall phenomenon. In the case of the hippocampus, it is important to know which regions send neural signals to which other regions. In the case of the hippocampus, the components are organized in a complex loop structure (see Figure 8.1). Furthermore, insofar as the hippocampus does not operate in isolation, investigators must relate the operations of the components of the hippocampus to other neural structures. Often computational modeling is required to evaluate whether a particular organization of parts carrying out proposed operations would be sufficient to realize the phenomenon (Rolls & Treves, 1998; McClelland, McNaughton & O'Reilly, 1995).

When mechanistic explanations appeal to the components of mechanisms to explain their behavior, they are clearly reductionistic. Moreover, the process of decomposition is iterative – the operation of a component part can itself be explained by another round of decomposition and localization. In fact, however, few mechanistic explanations involve more than two iterations of such decomposition. A major reason for this is that each decomposition addresses a different phenomenon (the operation of a component part as opposed to the operation of the whole mechanism). Once researchers have identified the parts of a mechanism and determined what operations they perform and how these operations are coordinated so as to enable the mechanism to realize the target phenomenon when in a given environment, the question that drove the inquiry has been answered.

3. Levels of Organization

Reduction is often characterized as appealing to lower levels. Despite frequent references to levels in discussions of reduction, what constitutes a level is often unspecified. In accounts of theory reduction, levels are often associated with broad scientific disciplines, so that one finds references to the level of physics, the level of psychology, etc., and to disciplines being reduced to disciplines at lower levels. Such accounts present a variety of problems. Physics deals with a broad range of entities, from the very small (subatomic particles) to the extremely large (galaxies). There doesn't seem to be a clear sense in which these reside at the same level. Nor is there an obvious sense in which the all the phenomena of physics lie at a lower level than those of, say, biology, which also deals with entities ranging from very small (viruses) to very large (ecosystems).

The division of inquiry into broad disciplines such as physics, biology, and psychology has much more to do with what humans are interested in studying – the behavior of ordinary physical objects (physical sciences), living organisms (biological sciences), behaving systems (behavioral sciences), and social activity (social sciences) – than with a hierarchy of levels of entities (Abrahamsen, 1987). As Abrahamsen notes, the phenomena studied by parts of physics also figure in the phenomena associated with life; accordingly, there are bridges between the physical and biological sciences (as well as between the other main divisions). Within these main divisions, differences in discipline may more closely correspond to a levels hierarchy. For example, molecular biology seems to be focused on a lower level than physiology, which deals with phenomena at a lower level than ecology. But even here relating disciplines to levels runs into difficulty. Microbiology and bacteriology seem to deal with different phenomena at the same level.

Attempts to sort out levels in terms of disciplines are fraught with problems (for further discussion, see Craver, forthcoming, ch. 5). A very different approach is to start not with the categorization provided by disciplines, but with phenomena in nature. An initially plausible view is to demarcate levels in terms of the size of the entities involved – small things are at a lower level than big things. This is the picture Churchland and Sejnowski (1988) adopt when they appeal to size scales to delineate levels of organization in the nervous system: molecules (1Å), synapses (1 μm), neurons (100 μm), networks (1 mm), maps (1 cm), and systems (10 cm). Wimsatt likewise proposes size as a way to differentiate levels, though he further elaborates the view by proposing that entities of the same size tend to "interact most strongly and frequently" (Wimsatt, 1976). Thus, levels are "*local maxima of regularity and predictability in the phase space of alternative modes of organization of matter*" (Wimsatt, 1994). Accordingly, he develops a stratified account according to which entities tend to fall into discrete clusters based on size: levels "*are constituted by families of entities usually of comparable size and dynamical properties, which characteristically interact primarily with one another, and which, taken together, give an apparent rough closure over a range of phenomena and regularities*" (Wimsatt, 1994; original emphasis).

If it were true that entities of a given size range tended to interact most frequently with other things of similar size, then this would be a principled approach. There are, however, plenty of examples of things of different sizes interacting causally. There are gravitational forces between very large objects (the earth) and very small objects (a molecule of hydrogen in the atmosphere). Storms can sweep seeds from one locale to another. Likewise, small things can causally affect big things – a bullet or a virus can kill an elephant. Absent a quantitative analysis, it is not clear that the claim that things tend to interact primarily with things of their own size is true.

Wimsatt often combines his analysis of levels by sizes with a compositional or mereological treatment of levels:

> By level of organization, I will mean here compositional levels – hierarchical divisions of stuff (paradigmatically but not necessarily material stuff) organized by part-whole relations, in which wholes at one level function as parts at the next (and at all higher) levels. (Wimsatt, 1994)

The first thing to note is that, although a consequence of this compositional analysis is that parts are smaller than the whole they constitute, this account is very different. It does not require that all parts of an entity must be of the same size – parts may vary radically as long as each is smaller than the whole. Second, the analysis only applies locally. Parts will be at a lower level than the whole to which they belong, but the compositional analysis does not tell us how to relate the parts to things outside the whole.

The mereological account of levels on its own, however, allows for arbitrary differentiation of the parts of a whole (see Craver, forthcoming, for a discussion of this and many other problems with formal treatments of mereology). Wedding the mereological account directly to mechanisms solves this problem. The component parts of a mechanism are the entities that perform the operations which together realize the phenomenon of interest. A structure within the mechanism may be well delineated (it has boundaries, continues to exist over time, is differentiated from the things around it, etc.). However, if it does not perform an operation that contributes to the realization of the phenomenon, it is not a working part of that mechanism. For example, while the gyri and sulci of the brain are well delineated, they are not working parts of the brain but byproducts of the way brains fold to conserve the length of axons (van Essen, 1997). The different working parts that constitute a level may be of

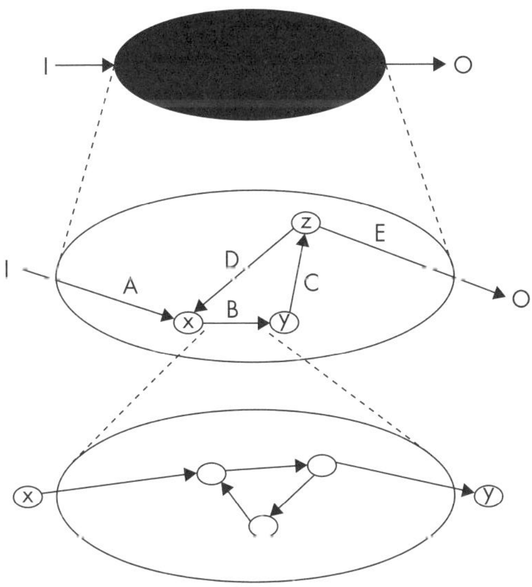

Figure 8.2 Moving down levels within a mechanism. The whole mechanism, shown in the top panel, is responsible for processing input I into output O. To explain how the mechanism is able to do this, investigators decompose it into its parts performing operations (indicated by uppercase letters) and determine how they produce changes in substrates (indicated by lowercase letters) in the system (middle panel). If one wants to explain how one component, B, performs its operation, a further round of decomposition into its parts and operations and determination of how they relate to one another is required (bottom panel).

different sizes – large parts, such as cell membranes, may interact with small parts such as individual sodium ions which are maintained in different concentrations on different sides of it.

To identify levels in terms of mechanisms, one starts with the mechanism identified in terms of the phenomenon for which it is responsible. The mechanism's working parts constitute the next lower level (see Figure 8.2). A consequence of this view is that levels are identified only with respect to a given mechanism; this approach does not support a conception of levels that extend across the natural world. Thus, it cannot address the question of whether glaciers, for example, are at the same level or at a higher or lower level than elephants, since they are not working parts of a common mechanism or related compositionally.

The account of levels within a mechanism can be generalized to multiple levels of organization once we recognize that a working part of a mechanism may itself be a mechanism. To explain how it performs its operation, investigators decompose the part into its own working parts. These parts are at a lower level than the component of the initial mechanism, and hence two levels below the mechanism as a whole. This process can clearly be iterated. The local nature of levels, however, emerges again as soon as we move down two levels. Because of the lack of a compositional relation between the sub-parts of one component of the mechanism and those of another component of that mechanism, the question of whether the sub-parts of two components are at the same level is not well defined.

While this approach cannot provide a global account of levels, it is sufficient for understanding the respects in which mechanistic explanation is reductionistic. The local character of the treatment of levels also has a rather surprising consequence that distinguishes mechanistic reduction from traditional views of reduction. Traditional views tend to assume that one can reduce higher-level explanations level by level until one reaches a fundamental level. On a theory reduction account, the theories at this level provide the foundation on which all higher-level theories are grounded. Even those who forego a theory-reduction perspective find it plausible that at some fundamental level we can identify the parts and operations out of which all higher-level mechanism are built. Theorists such as Kim (1998) then maintain that if we had a complete account of causal processes at this level, we would be able to determine all that happens in the universe. We would simply supply the initial conditions and make deductions from the laws governing the most basic level. Higher-level causal relations would overdetermine outcomes since these would already be determined at the lower level. But if the notion of levels is defined only locally, then on the mechanistic account we are not confronted by the prospect of a comprehensive lower level that is causally complete and closed. Such a picture of complete causal determination at a lower level is further brought into question when we consider why mechanistic explanations require relating levels of organization.

4. Relating Levels: Reduction and Autonomy

Although mechanistic explanation is reductive insofar as it appeals to the component parts and operations within a mechanism to explain the behavior of the mechanism, the reductive aspect alone is insufficient to explain the behavior of the mechanism. The parts of a mechanism behave in a particular way because of how they are organized in the mechanism. Information about how the parts are organized goes beyond the account of the parts and their operations. Moreover, the mechanism interacts causally with other entities. These interactions provide the input and set the conditions for the operation of the mechanism and information about them is not part of the reductive account characterizing the parts and operations within the mechanism. Securing information about both the organization within the mechanism and the relations between the mechanism and its environment requires going beyond the reductive aspect of mechanistic explanation and incorporating the results of other, autonomous inquiries.

This need to incorporate both a reductive component and information secured from autonomous inquires at higher levels can be illustrated in a simple biological example. To produce your favorite ale, a brewer brings together the ingredients necessary for fermentation and creates the proper condition for fermentation to occur. At a lower level, individual yeast cells take in glucose and generate alcohol along with carbon dioxide. At a yet lower level, enzymes such as glucose-6-phosphatase in the yeast's cytoplasm catalyze specific reactions in the biochemical pathway from glucose to alcohol. The operation of the enzymes in catalyzing a specific reaction depends upon the context in the yeast cell, and that cell's situation in the brewer's vats. The enzymes in yeast remain able to catalyze reactions, but if no glucose is available or the brewer does not create the proper conditions, fermentation will not occur. These conditions are additional to providing the enzymes and the inquiry into which conditions are best suited for fermentation is distinct from the inquiry into the operation of the enzymes. Indeed, knowledge of how to set up conditions for fermentation was acquired by brewers long before the development of biochemistry in the early twentieth century,[2] and that knowledge was not supplanted by the investigations of biochemists. On the contrary, biochemists employed such knowledge in setting up the experimental conditions in which they could study the operation of the enzymes.

A similar relationship between reductive appeals to the parts of a mechanism and autonomous investigations of the organization of those components and of how the mechanism is situated in an environment is found in the case of the primate visual system. Reductive investigations in the second half of the twentieth century identified the different brain areas that comprise the visual system and the operation of each in extracting information from visual inputs (van Essen & Gallant, 1994). Such inquiry determined that V1 detects edges (amongst other things), V4 computes color constancy and identifies shape, and MT computes perceived motion. None of these operations is itself the activity of seeing an object. Although one might be tempted to identify seeing objects with the firing of cells in inferotemporal cortex, where the process of computing the identity of the object is thought to occur, this temptation should be resisted. Inferotemporal cortex does not operate in isolation, but only in conjunction with other components of the visual system. It simply represents an output component of a complex mechanism. Each of the various components of the mechanism carry out operations that together realize the phenomenon of perceiving objects both as entities of a particular kind and also as having a given shape or a certain color. Moreover, seeing only occurs when all the parts are organized in the right way and the whole organism engages a visual world. A person recognizes a dog when her retinal cells are activated by patterns of light reaching them from a dog in the environment, the components of her visual system are properly connected, and other brain processes, including those required for attention, operate normally. These other processes, some of them involving mechanisms themselves, produce the conditions for the phenomenon in question.

Understanding how the visual system is organized, coordinated with other physiological systems, and responsive to external stimuli requires knowledge beyond the specification of the parts of the visual system and their operation. Moreover, investigations of these relations can determine important regularities that are not provided by the reductive inquiry. For example, psychophysicists began to identify regularities about how our visual system responds to sensory stimuli almost 150 years ago (Fechner, 1860; Stevens, 1957), and these investigators did not know about the parts of the system or how those parts operated. James Gibson (1966, 1979) and subsequent ecological psychologists identified many sorts of information to which the visual system is responsive without engaging in any study of the components of the system. Moreover, these researchers have produced knowledge that could not have been acquired by such reductive inquiry. Psychophysicists and ecological psychologists complement the reductive inquiries of neurophysiologists and have not been rendered unnecessary by the neurophysiologists' success.[3]

Not only can one study the performance of a mechanism without knowing its component parts and their operations, but what the mechanism as a whole does is typically quite different than the operations performed by its parts. Neurons in different parts of the visual system generate action potentials in response to release of transmitters by cells on which they synapse, for example, while the (ventral) visual system as a whole identifies what object is presented to the person and makes that information available to other cognitive systems (those engaged, for example, in encoding memories or making decisions). As this illustration also makes clear, the mechanism as a whole may in fact constitute a component of a larger mechanism that does something still different (enabling the organism to act). The information processing of neural ensembles is different from the production of action potentials in individual cells.

The fact that mechanisms perform different activities than do their parts manifests itself in the fact that the activities of whole mechanisms are typically described in different vocabulary than are component operations. Traditional accounts of theory reduction implicitly recognized this fact by requiring bridge principles to connect the different vocabularies used in different sciences, but little notice was given as to why different sciences employ different vocabularies. The vocabulary used in each science describes different types of entities and different operations – one describes the parts and what they do, whereas another describes the whole system and what it does. Relating the vocabulary used in the different sciences requires consideration of the compositional relations between the entities and their operations. The substantive knowledge required to establish these relations is not derived from the lower-level laws but requires additional empirical investigation. Recognizing this reveals that even the theory-reduction account must incorporate higher-level knowledge, and so is not as reductive as it appears.

Throughout this discussion I have been making reference to the way the components of a mechanism are organized. It is the fact that mechanisms organize the operations performed by the parts that enables them to do things no part alone can do, which in turn requires higher-level inquiries into how the mechanism engages its environment.

Organization itself is not something inherent in the parts (even self-organizing systems only organize themselves under appropriate conditions). Accordingly, investigators who already understand in detail how the parts behave are often surprised by what happens when they are organized in particular ways. To appreciate this, consider engineering design. The primary activity of a designer is to put components together in novel ways to produce new activities. A designer does not develop a new mechanism by creating it de novo, but by imposing novel organization on components that already exist. An indication of the critical contribution of organization is the fact that engineers who are successful in discovering organizational principles that enable mechanisms to perform activities are able to secure patents for their design are rewarded financially or with fame.

In virtue of being organized systems, mechanisms do things beyond what their components do. But beyond this, the organization of the components typically integrates them into an entity that has an identity of its own. As a result, organized mechanisms become the focus of relatively autonomous disciplines – disciplines which deploy their own tools of investigation and develop their own distinctive accounts of the phenomena associated with these mechanisms. Thus, to understand the autonomy achieved by higher levels within a mechanistic framework, we need to focus on the sorts of organization that figure in mechanisms.

5. Types of Organization and the Generation of Higher Levels

By introducing the notion of *aggregativity*, Wimsatt (1986) provided a baseline account of collectives that lack organization. In an aggregate, such as a pile of sand, the component parts are simply amassed together without any specific organization. The components each behave as they would outside of the aggregate; specifically, they do not interact in ways that result from specific dependencies of one part on another. Parts can be substituted for one another and, as new parts are added to the aggregate, the behavior of the whole depends simply on the number of parts present (at least until critical points are reached where interesting non-linear interactions such as landslides begin to occur). As a consequence, aggregates do not produce entities at higher levels or require new inquiries.

A first step away from aggregativity is for some parts to depend on the prior operation of others in order to perform their own operations. In a linear organization (of which human assembly lines provide exemplars), the product of the operation of one part is operated on by the next in a sequence until an output product is produced. A system organized in this way can accomplish more than can any given component. Whereas in an aggregate, any part can replace another, linearly organized systems depend on having the right order of parts so that the products of one operation are made available as inputs to a subsequent operation. Nonetheless, the organization is still essentially additive: the operation of one component is simply added to the operation of another component. Understanding linearly organized systems requires only modest additions to the knowledge of how the components work.

Engineers often begin by designing linear systems and scientists begin by proposing linear models for natural phenomena. One reason is that our conscious thinking is sequential – we think one thought and then another. So when we try to understand a system, we conceptualize one operation followed by another. (Even when processes are known not to be strictly linear – when they depend upon feedback loops that connect operations earlier in the sequence to those later in the sequence – scientists often represent them linearly. Thus, despite the fact that some of the reactions in fermentation are linked with ones earlier or later in the pathway, the fermentation pathway is commonly represented as a linear sequence of reactions.) But scientists, especially in the life sciences, have repeatedly discovered that natural systems are not organized linearly but exhibit various types of interactivity such as cycles. In cycles, the product of a sequence of operations feeds back into an earlier step in the process. We have already encountered one example: the cyclic pathways linking the various regions of the hippocampus (see Figure 8.1). The Krebs cycle in oxidative metabolism is another clear and well-known exemplar of cyclic organization – the final product of a series of oxidation reactions, oxaloacetic acid, is combined with additional acetyl-CoA arriving from carbohydrate, fat, or protein metabolism, to produce citric acid. This is then oxidized in another round of the cycle. Biochemists, physiologists, and ecologists have discovered a plethora of such cycles.

Cycles often allow the results of an operation to feed back on earlier operations and regulate them. Figure 8.3 shows a negative feedback

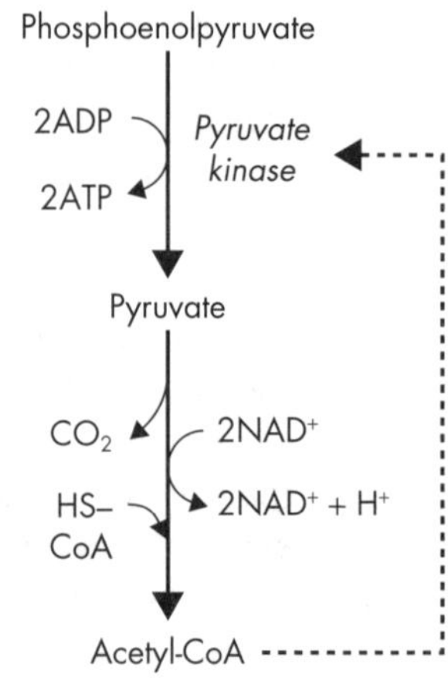

Figure 8.3 Feedback loop (dotted line) at the point of entry to the Krebs cycle. If acetyl-CoA builds up from not being metabolized by the Krebs cycle, it feeds back onto the dephosphorylation of phosphoenolpyruvate, causing that reaction to halt.

loop at the entry point to the Krebs cycle where accumulation of acetyl-CoA serves to inhibit the earlier formation of pyruvate from phosphoenolpyruvate, thereby preventing additional accumulation of acetyl-CoA when it is not needed. The usefulness of negative feedback organization seems to have first been discovered in the third century BCE by Ktesibios in his design of the water clock. In order for water to flow at a regular rate into the vessel measuring time, a constant volume of water in the supply tank was required. Ktesibios achieved this by inserting a float where water entered the supply tank so that whenever the water exceeded that height, the float would block the entrance, preventing any more water from entering. The idea of negative feedback, however, was not easily generalized to other contexts but had to be continuously rediscovered (Mayr, 1970). Watt's introduction of the centrifugal governor for the steam engine, and Maxwell's (1868) mathematical analysis of it, led finally to recognition of negative feedback as a general technique for regulating behavior of mechanisms. In the mid-twentieth century the cyberneticists celebrated negative feedback and promoted cyclic design as a fundamental principle (Wiener, 1948) in the organization of biological and social systems. From the point of view of understanding a mechanism, when negative feedback is employed, the operation of some components falls under the influence of other components, and a theorist cannot account for the behavior of the whole system by just adding together the outputs of the component parts. Rather, the theorist must appreciate how the components are constrained by the conditions created by the operation of other components.

Already with a simple mechanism that performs a function different than the operations of its parts we had introduced a higher level of organization that exhibited a kind of autonomy from inquiries at the lower level. The higher level investigated the engagement of the mechanism as a whole with its environment. But as more complex modes of organization are introduced, this autonomy grows. When designs such as negative feedback are introduced, the system develops a kind of insulation from certain perturbations in the environment. Understanding the behavior of the mechanism requires not just knowing its parts and operations but the capacities provided by the negative feedback.

Although negative feedback is now reasonably well known, positive feedback, in which the products of two operations each facilitate the other, often receives less attention. This is due to the fact that positive feedback in many situations leads to out of control, runaway behavior. But in some contexts such interactions are a powerful force for developing higher-level structure. Positive feedback can provide a basis for self-organization – the ability of a set of components to organize themselves into structures which perform operations beyond what the components themselves are capable of (Kaufmann, 1993). Here I will limit my focus to positive feedback within networks of already organized components.

For the most part, human thinking about networks has started from one of two designs – regular lattices in which components are connected to those closest to them and randomly connected networks (investigated by Erdös and Rényi) in which each component has a random probability of being linked directly to any other unit. But two developments in thinking about network design in the late twentieth century revealed forms of organization in networks that result in far more interesting sorts of system behavior. Inspired by Stanley Milgram's 1967 letter-mailing experiment and John Guare's 1990 play, *Six Degrees of Separation*, Duncan Watts (Watts & Strogratz, 1998) explored what he termed *small worlds.* These are networks in which there is high local connectivity, as in regular lattices, supplemented by a few long-range connections. These long-range connections serve to bring the path-length between any two nodes into the same range as in a randomly connected network. Such networks show how one can have systems that are largely modular in design (clusters of units are primarily connected with each other and can operate as a higher-level unit) without the modules becoming completely isolated from one another, as in Fodor's (1983) account of mental modules. A few long-range connections enable these modules to coordinate their activity, and thus function as integrated systems at yet higher levels of organization.

The second new idea was that nodes in networks might have very different numbers of connections to other nodes, some being connected to just one or two, and a very few to a very large number. Barabási and Albert (1999) characterize networks in which the connectivity of nodes drops off according to a power-law as *scale-free*. In scale-free networks many components of a network can be destroyed while the rest of the network retains its integrity and continues to function. Many naturally occurring networks, such as metabolic systems in cells and human social networks, exhibit the properties of small worlds and scale-freeness. The ubiquity of such networks raises the question of what additional fruitful properties result from such designs that we have previously failed to appreciate.

One can gain further appreciation of the importance of nonlinear modes of organization in networks of components by considering the recent history of research on artificial neural networks. The discovery of the backpropagation training algorithm for multilayer networks in the mid-1980s (Rumelhart, Hinton, & Williams, 1986) focused a great deal of attention on feedforward networks in which activity in one layer of units contributed to activity of units in the next layer. These networks operate in a sequential manner.[4] But other researchers were already concentrating on far more interactive networks (Hinton & Sejnowski, 1986). One simple addition to a feedforward network is to employ the activity pattern on later units as additional inputs to earlier layers of units in subsequent epochs, resulting in what is called a *simple recurrent network*. Elman (1991, 1993) showed that such networks could learn to process complex grammatical forms involving the type of long-range dependencies found in natural languages without invoking internal representations of linguistic structure. Other network modelers have used positive feedback between units in networks to capture some of the rich dynamics of human behavior (van Leeuwen, Steyvers, & Nooter, 1997). Recognition of the fact that many real-world networks employ small-world and scale-free features has led investigators to explore their use in neural network modeling (Gong & van Leeuwen, 2003, 2004). A general feature of such research is to reveal how networks of simple processing units organized in more complex ways could generate behavior that initially seemed quite beyond the capacity of such networks.

My interest in this section is in how organization enables systems of components to exhibit behavior different in character than that exhibited by the components. Such organized systems become the focus of their inquiry that is autonomous from inquiry into the behavior of the components and focuses on how these systems engage the world in their own way. Herbert Simon drew attention to this in his analysis of hierarchically organized complex systems (Simon, 1962). A century earlier Claude Bernard (1865) investigated the ability of living systems to maintain themselves through a wide-range of environment changes and introduced the important distinction between the internal environment of an organism and its external environment. He further articulated the idea that individual parts of organisms perform operations needed to maintain the constancy of the internal environment. This idea was partially explicated with Cannon's (1929) account of mechanisms of homeostasis and the cyberneticist's account of negative feedback. But a critical feature of complex systems to which Simon drew attention was the interface – the boundary between a system and its environment. In biological systems, membranes provide a means by which organisms (including single-cell organisms) can control admission of foodstuffs and other needed materials from their environments and expel toxins back into the environment. Maintenance of membranes does not come for free – organisms must build and maintain their own membranes, drawing on energy and resources from their environment. Accordingly two of the critical components that Tibor Gánti (1975, 2003) included in his conception of a chemoton, the simplest chemical system that he maintained would exhibit the properties of life, were a metabolic system and membranes, with the membrane segregating the metabolic system from the environment and controlling access to it. (The third was a control system, which I will not consider here.) At the center of Gánti's conception of the chemoton are cyclic processes which enable the chemoton to maintain itself and even build more of its metabolic components and its membrane (relying on physicochemical processes to divide into daughter chemotons).

The important feature of such systems as Gánti's chemoton is that they are able to maintain themselves in relative independence of their local environments. To capture this feature of living systems, Alvaro Moreno and his colleagues speak of them as *autonomous*,[5] and characterize an autonomous system as

> a far-from-equilibrium system that constitutes and maintains itself establishing an organizational identity of its own, a functionally integrated (homeostatic and active) unit based on a set of endergonic–exergonic couplings between internal self-constructing processes, as well as with other processes of interaction with its environment. (Ruiz-Mirazo, Peretó, & Moreno, 2004)

Although the notion of an autonomous system is different from the autonomy of different inquiries, which has been my focus here, they are related. The fact that biological systems are autonomous is part of what motivates inquiries into their behavior and the organizational conditions that give rise to it that is distinct from inquiries into their component parts and their operations. The relationship, however, is the same as in mechanisms generally: the parts, which are the focus of reductive inquiry, are organized and situated in an environment. Understanding how they are organized and situated in not redundant to what the reductive inquiry reveals. It rather provides additional information that is required to understand the mechanism. The recognition that biological systems are autonomous systems simply makes the need for these inquiries focused on higher levels all the more important.

6. Conclusion

Mechanistic explanation is reductionistic insofar as it appeals to the components of a mechanism to explain its activity. But insofar as the phenomenon generated by a mechanism depends upon the organization of the parts and the conditions impinging upon the mechanism from without, mechanistic explanation also recognizes the autonomy of higher-level investigations. Modes of organization are not determined by the components but are imposed on them. (This is true even when systems self-organize – once they have organized, the components are subject to the constraints imposed by the organization as a whole.) The contribution of organization in creating mechanisms that do things their parts cannot do undergirds the need for scientists to discover the particular forms of organization realized in a mechanism. This is what higher-level disciplines, such as psychology, have the resources to provide. Their autonomous contribution is secure even if higher-level activities are not multiply realizable. At the same time, mechanistic explanation also recognizes the value of reductionistic investigations into how the components perform their operations. Higher-level inquiries and reductionistic inquiries complement each other, and often provide heuristic guidance to each other. Neither on its own suffices and neither can do the work of the other.

Acknowledgments

I thank Cory Wright for productive discussion of points in this chapter and for presenting an earlier version of it for me at the 2004 meeting of the Society for Philosophy and Psychology in Barcelona, Spain. I also thank the participants and audience at a symposium of the Boston University Center for Philosophy and History of Science in May, 2005, during which a version of this chapter was also presented, for their helpful questions and discussion.

Notes

1 Although some philosophers speak of reducing psychological processes to neural ones, the relation between the components identified in psychology and those identified in neuroscience is actually identity between the component performing a psychological operation and a neurally identified brain part. It is important to recognize that these identity claims are often advanced at the outset of research as hypotheses to guide further inquiry. Accordingly, McCauley and I speak of the *heuristic identity theory* (Bechtel & McCauley, 1999; McCauley, in press). This addresses an objection raised to traditional identity theories (Smart, 1959) that evidence could never support psycho-physical identity claims as opposed to psycho-physical correlations. Accepting an identity hypothesis commits a researcher to a host of further consequences. If these consequences are borne out, the research program will have incorporated the identity claim

at its foundation and researchers will not be asking whether a mere correlational claim could serve as well.

2 Indeed, physiological investigations of yeast were conducted while it was still unknown what happened inside them. Pasteur (1860), for example, conducted detailed studies of the conditions in which fermentation would occur and established the important result that yeast carry out fermentation when in an oxygen-free environment, an effect known as the Pasteur effect. But Pasteur himself rejected the prospects of a chemical account of the reactions occurring within yeast cells.

3 Adopting this perspective, we can recognize the mistake of eliminative materialists in maintaining that knowledge of the brain will eliminate the need for folk psychology. Folk psychology, like social psychology, characterizes regularities in the way cognitive agents respond to situations arising in their environment (Bechtel & Abrahamsen, 1993). This is not information that neuroscientists themselves are interested in or have the tools to procure (but see Bickle, Chapter 12 in this volume, for an opposing view).

4 Backpropagation, as the name suggests, already uses backwards projections from output units to weights in the network, but only in the process of training the network, not when the network is solving a particular problem.

5 The notion of autonomy has also been developed by Varela (1979; see also Maturana & Varela, 1980). Varela, however, does not invoke thermodynamic considerations in his account of autonomy. It is the thermodynamic phenomenon of being far from equilibrium which imposes on living systems the need to recruit resources and energy from their environment and to use these to maintain themselves. It is this feature that makes autonomy so critical to understanding biological mechanisms.

References

Abrahamsen, A. A. (1987). Bridging boundaries versus breaking boundaries: Psycholinguistics in perspective. *Synthese*, *72*, 355–388.

Atkinson, R. C., & Shiffrin, R. M. (1968). Human memory: A proposed system and its control processes. In K. W. Spence & J. T. Spence (Eds.), *The Psychology of Learning and Motivation: Advances in Research and Theory* (Vol. 2, pp. 89–195). New York: Academic.

Barabási, A.-L., & Albert, R. (1999). Emergence of scaling in random networks. *Science*, *286*, 509–512.

Bechtel, W. (2005). The challenge of characterizing operations in the mechanisms underlying behavior. *Journal of the Experimental Analysis of Behavior*, *84*, 313–325.

Bechtel, W., & Abrahamsen, A. (1993). Connectionism and the future of folk psychology. In S. M. Christensen & D. R. Turner (Eds.), *Folk Psychology and the Philosophy of Mind* (pp. 341–367). Hillsdale, NJ: Lawrence Erlbaum, and in R. Burton (Ed.), *Minds: Natural and Artificial* (pp. 69–100). Albany, NY: SUNY Press.

Bechtel, W., & Abrahamsen, A. (2005). Explanation: A mechanist alternative. *Studies in History and Philosophy of Biological and Biomedical Sciences, 36*, 421–441.

Bechtel, W., & Abrahamsen, A. (in press). Phenomena and mechanisms: Putting the symbolic, connectionist, and dynamical systems debate in broader perspective. In R. Stainton (Ed.), *Contemporary Debates in Cognitive Science*. Oxford: Blackwell.

Bechtel, W., & Hamilton, A. (in press). Reduction, integration, and the unity of science: Natural, behavioral, and social sciences and the humanities. In T. Kuipers (Ed.), *Philosophy of Science: Focal Issues*. New York: Elsevier.

Bechtel, W., & McCauley, R. N. (1999). Heuristic identity theory (or back to the future): The mind-body problem against the background of research strategies in cognitive neuroscience. In M. Hahn & S. C. Stoness (Eds.), *Proceedings of the 21st Annual Meeting of the Cognitive Science Society* (pp. 67–72). Mahwah, NJ: Lawrence Erlbaum Associates.

Bechtel, W., & Mundale, J. (1999). Multiple realizability revisited: Linking cognitive and neural states. *Philosophy of Science, 66*, 175–207.

Bechtel, W., & Richardson, R. C. (1993). *Discovering Complexity: Decomposition and Localization as Strategies in Scientific Research*. Princeton, NJ: Princeton University Press.

Bernard, C. (1865). *An Introduction to the Study of Experimental Medicine*. New York: Dover.

Bickle, J. (1998). *Psychoneural Reduction: The New Wave*. Cambridge, MA: MIT Press.

Bickle, J. (2003). *Philosophy and Neuroscience: A Ruthlessly Reductive Account*. Dordrecht: Kluwer Academic Publishers.

Cannon, W. B. (1929). Organization of physiological homeostasis. *Physiological Reviews, 9*, 399–431.

Churchland, P. S., & Sejnowski, T. J. (1988). Perspectives on cognitive neuroscience. *Science, 242*, 741–745.

Craver, C. (forthcoming). *Explaining the Brain: What a Science of the Mind–Brain Could Be*. New York: Oxford University Press.

Cummins, R. (2000). "How does it work?" versus "What are the laws?": Two conceptions of psychological explanation. In F. Keil & R. Wilson (Eds.), *Explanation and Cognition* (pp. 117–144). Cambridge, MA: MIT Press.

Elman, J. L. (1991). Finding structure in time. *Cognitive Science, 14*, 179–211.

Elman, J. L. (1993). Learning and development in neural networks: The importance of starting small. *Cognition*, *48*, 71–99.

Fechner, G. T. (1860). *Elemente der Psychophysik*. Leipzig: Breitkopf & Härtel.

Fodor, J. A. (1974). Special sciences (or: the disunity of science as a working hypothesis). *Synthese*, *28*, 97–115. Reprinted in N. Block (Ed.), *Readings in Philosophy of Psychology* (Vol. 1, pp. 120–133). Cambridge, MA: Harvard University Press, 1980.

Fodor, J. A. (1983). *The Modularity of Mind*. Cambridge, MA: MIT Press.

Gánti, T. (1975). Organization of chemical reactions into dividing and metabolizing units: The chemotons. *BioSystems*, *7*, 15–21.

Gánti, T. (2003). *The Principles of Life*. New York: Oxford University Press.

Gibson, J. J. (1966). *The Senses Considered as Perceptual Systems*. Boston: Houghton Mifflin.

Gibson, J. J. (1979). *The Ecological Approach to Visual Perception*. Boston: Houghton Mifflin.

Glennan, S. (1996). Mechanisms and the nature of causation. *Erkenntnis*, *44*, 50–71.

Glennan, S. (2002). Rethinking mechanistic explanation. *Philosophy of Science*, *69*, S342–S353.

Gong, P., & van Leeuwen, C. (2003). Emergence of a scale-free network with chaotic units. *Physica A: Statistical Mechanics and its Applications*, *321*, 679–688.

Gong, P., & van Leeuwen, C. (2004). Evolution to a small-world network with chaotic units. *Europhysics Letters*, *67*, 328–333.

Hinton, G., & Sejnowski, T. J. (1986). Learning and relearning in Boltzmann machines. In D. E. Rumelhart & J. L. McClelland (Eds.), *Parallel Distributed Processing: Explorations in the Microstructure of Cognition* (Vol. 1, pp. 282–317). Cambridge, MA: MIT Press.

Kaufmann, S. A. (1993). *The Origins of Order*. Oxford: Oxford University Press.

Kim, J. (1998). *Mind in a Physical World: An Essay on the Mind–Body Problem and Mental Causation*. Cambridge, MA: MIT Press.

Kolers, P. A., & Roediger, H. L. (1984). Procedures of mind. *Journal of Verbal Learning and Verbal Behavior*, *23*, 425–449.

Machamer, P., Darden, L., & Craver, C. (2000). Thinking about mechanisms. *Philosophy of Science*, *67*, 1–25.

Maturana, H. R., & Varela, F. J. (1980). Autopoiesis: The organization of the living. In H. R. Maturana & F. J. Varela (Eds.), *Autopoiesis and Cognition: The Realization of the Living* (pp. 59–138). Dordrecht: D. Reidel.

Maxwell, J. C. (1868). On governors. *Proceedings of the Royal Society of London*, *16*, 270–283.

Mayr, O. (1970). *The Origins of Feedback Control.* Cambridge, MA: MIT Press.

McCauley, R. N. (in press). Reduction: Models of cross-scientific relations and their implications for the psychology–neuroscience interface. In P. Thagard (Ed.), *Handbook of the Philosophy of Psychology and Cognitive Science.* Amsterdam: Elsevier.

McClelland, J. L., McNaughton, B., & O'Reilly, R. C. (1995). Why there are complementary learning systems in the hippocampus and neocortex: Insights from the successes and failures of connectionist models of learning and memory. *Psychological Review*, *102*(3), 419–457.

Nagel, E. (1961). *The Structure of Science: Problems in the Logic of Scientific Explanation.* New York: Harcourt, Brace, and World.

Neisser, U. (1967). *Cognitive Psychology.* New York: Appleton-Century-Crofts.

Oppenheim, P., Putnam, H. (1958). The unity of science as a working hypothesis. In H. Feigl & G. Maxwell (Eds.), *Concepts, Theories, and the Mind–Body Problem* (pp. 3–36). Minneapolis: University of Minnesota Press.

Pasteur, L. (1860). Mémoire sur la fermentation alcoolique. *Annales de Chimie, 3è Sér.*, *58*, 323–426.

Polger, T. (2004). *Natural Minds.* Cambridge, MA: MIT Press.

Posner, M. I. (1978). *Chronometric Explorations of Mind.* Hillsdale, NJ: Lawrence Erlbaum Associates.

Putnam, H. (1967). Psychological predicates. In W. H. Capitan & D. D. Merrill (Eds.), *Art, Mind and Religion* (pp. 37–48). Pittsburgh: University of Pittsburgh Press.

Redish, A. D. (1999). *Beyond the Cognitive Map.* Cambridge, MA: MIT Press.

Roediger, H. L., Buckner, R. L., & McDermott, K. B. (1999). Components of processing. In J. K. Foster & M. Jelicic (Eds.), *Memory: Systems, Process, or Function* (pp. 32–65). Oxford: Oxford University Press.

Rolls, E. T., & Treves, A. (1998). *Neural Networks and Brain Function.* Oxford: Oxford University Press.

Ruiz-Mirazo, K., Peretó, J., & Moreno, A. (2004). A universal definition of life: Autonomy and open-ended evolution. *Origins of Life and Evolution of the Biosphere*, *34*, 323–346.

Rumelhart, D. E., Hinton, G. E., & Williams, R. J. (1986). Learning internal representations by error propagation. In D. E. Rumelhart & J. L. McClelland (Eds.), *Parallel Distributed Processing: Explorations in the Microstructure of Cognition* (Vol. 1, pp. 318–362). Cambridge, MA: MIT Press.

Scoville, W. B., & Milner, B. (1957). Loss of recent memory after bilateral hippocampal lesions. *Journal of Neurology, Neurosurgery, and Psychiatry*, *20*, 11–21.

Shapiro, L. (2004). *The Mind Incarnate.* Cambridge, MA: MIT Press.

Simon, H. A. (1962). The architecture of complexity: hierarchic systems. *Proceedings of the American Philosophical Society, 106*, 467–482.

Smart, J. J. C. (1959). Sensations and brain processes. *Philosophical Review, 68*, 141–156.

Stevens, S. S. (1957). On the psychophysical law. *Psychological Review, 64*, 153–181.

Tulving, E. (1983). *Elements of Episodic Memory.* New York: Oxford University Press.

van Essen, D. C. (1997). A tension-based theory of morphogenesis and compact wiring in the central nervous system. *Nature, 385*, 313–318.

van Essen, D. C., & Gallant, J. L. (1994). Neural mechanisms of form and motion processing in the primate visual system. *Neuron, 13*, 1–10.

van Leeuwen, C., Steyvers, M., & Nooter, M. (1997). Stability and intermittency in large-scale coupled oscillator models for perceptual segmentation. *Journal of Mathematical Psychology, 41*, 319–344.

Varela, F. J. (1979). *Principles of Biological Autonomy.* New York: Elsevier.

Watts, D., & Strogratz, S. (1998). Collective dynamics of small worlds. *Nature, 393*, 440–442.

Wiener, N. (1948). *Cybernetics: Or, Control and Communication in the Animal Machine.* New York: John Wiley.

Wimsatt, W. C. (1976). Reductionism, levels of organization, and the mind–body problem. In G. Globus, G. Maxwell, & I. Savodnik (Eds.), *Consciousness and the Brain: A Scientific and Philosophical Inquiry* (pp. 202–267). New York: Plenum.

Wimsatt, W. C. (1986). Forms of aggregativity. In A. Donagan, N. Perovich, & M. Wedin (Eds.), *Human Nature and Natural Knowledge* (pp. 259–293). Dordrecht: Reidel.

Wimsatt, W. C. (1994). The ontology of complex systems: Levels, perspectives, and causal thickets. *Canadian Journal of Philosophy, Supplemental Volume 20*, 207–274.

Wright, C., & Bechtel, W. (in press). Mechanisms and psychological explanation. In P. Thagard (Ed.), *Philosophy of Psychology and Cognitive Science.* Elsevier.

9

ENRICHING PHILOSOPHICAL MODELS OF CROSS-SCIENTIFIC RELATIONS: INCORPORATING DIACHRONIC THEORIES

Robert N. McCauley

1. Introduction: Simple Reduction and Beyond

Traditional and New Wave models of reduction in science have not lacked for ambition. Philosophers have presented single models to account for the full range of interesting intertheoretic relations, for scientific progress, and for the unity of science (Nagel, 1961; Oppenheim & Putnam, 1958). Early critics attacked the logical empiricists' proposals about the character of intertheoretic connections (Feyerabend, 1962; Kuhn, 1962). New Wave reductionists have similarly argued that various intertheoretic relations fall at different points on a continuum of goodness-of-intertheoretic-mapping. Still, whatever their differences with the logical empiricists, New Wave reductionists have retained traditional aspirations for a single comprehensive model of reduction that will make sense of the wide range of intertheoretic relations, of progressive scientific change, and of how the various sciences hang together (Hooker, 1981; Churchland & Churchland, 1990; Bickle, 1998). Both logical empiricists and their New Wave successors proffer such unified, multipurpose models.

Regardless of the field, multipurpose tools typically sacrifice precision for versatility. Recent analyses of mechanistic explanation have helped to reveal that these models of scientific reduction are no exceptions to this rule, and the cost of sacrificing precision is one that the mechanists are unwilling to pay. Traditional and New Wave reductionists manifest allegiances

(1) to the (virtually exclusive) analysis of theories and intertheoretic relations and

(2) to conceptions of explanatory levels in science rooted in considerations pertaining to the size of and the mereological relations between the sciences' objects of study. (Discussions of both explanatory levels and mereological relations follow in subsequent sections.)

By contrast, advocates of mechanistic analysis offer detailed accounts of particular systems' functioning that survey their components, operations, and the larger systems to which they contribute (Bechtel & Richardson, 1993; Glennan, 1996; Machamer, Darden, & Craver, 2000; Craver, 2001; Bechtel, 2006). The mechanists' approach to philosophical questions concerning cross-scientific relations is primarily data-driven from the bottom up.

Explanatory pluralism (McCauley, 1986, 1996, in press; McCauley & Bechtel, 2001) also eschews a single multipurpose model of reduction, arguing that considerations and dynamics bearing on theory succession within particular sciences often substantially diverge from those pertaining to cross-scientific relations. The explanatory pluralist welcomes the detailed analyses of specific mechanisms in nature that the mechanists are furnishing as a pivotal contribution to our understanding of scientific explanation in cross-scientific contexts, but surveys additional issues as well. Explanatory pluralism also investigates, across explanatory levels, such matters as borrowing concepts and theoretical strategies, sharing experimental techniques and tools, and, perhaps most importantly, tapping new sources of evidence, especially experimental evidence. Explanatory pluralism emphasizes that because just about everything about science is so hard to do, cross-scientific opportunism is plenteous (McCauley, 2000; in press).

Explanatory pluralism explores a broader array of considerations and envisions a more wide-ranging framework for the analysis of cross-scientific relations than do mechanistic analyses. Nevertheless, the two approaches are not at odds. They reject both traditional and New Wave reductionists' absorption with laws and theories and the sufficiency of mereological considerations to distinguish levels of explanation plausibly (items (1) and (2) above). Explanatory pluralism, however, also retains a concern for more traditional (top-down) reflection on the sciences' overall arrangement. Ideally, mechanistic analysis contributes to filling out the explanatory pluralist picture and provides a means for checking explanatory pluralism's more general declarations about explanatory levels in science. In the light of both mechanists' findings about the various cases they examine and a more fine-grained treatment of the wide variety of cross-scientific relations, explanatory pluralists aim to refine our large-scale conceptions of explanatory levels in science and of how they are connected. By laying out a hitherto mostly neglected distinction between theories and investigations of synchronic, structural phenomena, on the one hand, and of diachronic phenomena (especially larger-scale and longer-term diachronic phenomena), on the other, this chapter aims to advance that "more fine-grained treatment of the wide variety of cross-scientific relations" a bit further.

Section 2 reviews the explanatory pluralist's principal arguments to date against all-purpose models of scientific reduction first and then briefly sketches the resulting view of cross-scientific relations that explanatory pluralism has supplied heretofore. Section 3 enlarges this view, taking inspiration from a distinction between two types of inquiry among the sciences, viz., between the study of natural structures without or at least with little reference to time and the study of natural processes and changes in nature over time. Up to now, philosophical discussions concerning cross-scientific settings have focused almost exclusively on the former. Section 4 explores the more complicated portraits of (1) analytical levels in science, (2) cross-scientific relations (offering a quick illustration from the cognitive sciences of prominent types of intertheoretic and cross-scientific connections that can arise), and (3) reduction.

2. Some Familiar Reasons why Reduction in Science Is Not Simple: Distinguishing between Intralevel and Interlevel Relations

William Wimsatt (1976) first suggested that behind Ernest Nagel's traditional conception of reduction in science lurked two quite different sorts of intertheoretic relations. Distinguishing between *intralevel* or *successional* reduction (concerned with progress over time within some science) and *interlevel* reduction (concerned with what I have been referring to as cross-scientific relations), Wimsatt maintains that philosophers and scientists (especially those interested in the relations of psychology and neuroscience) frequently confuse their diverging dynamics (see Figure 9.1). He has eliminative materialism, in particular, in his sights. Eliminative materialists draw inferences about the elimination of psychology and its objects of study such as beliefs,

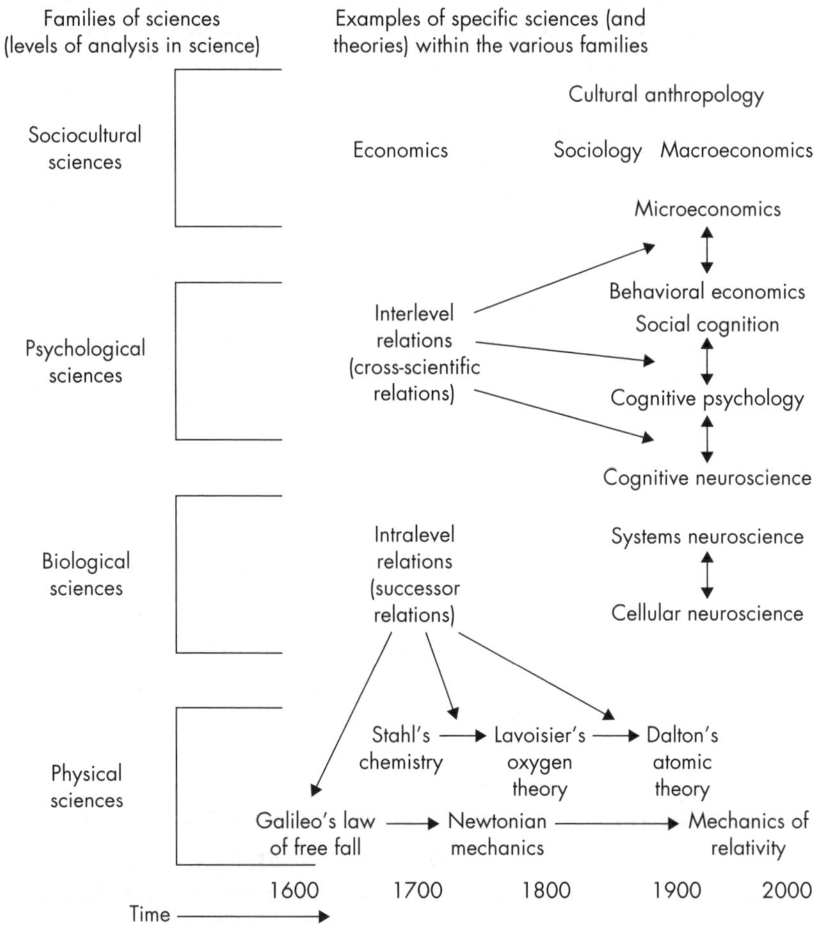

Figure 9.1 Distinguishing intralevel and interlevel relations

hopes, and other mental states and, ultimately, the mind itself) on the basis of projections about the emergence of superior neuroscientific accounts of the phenomena to be explained. Expanding on Wimsatt's objection, I (McCauley, 1986) argued that the problem is that eliminativists come to these conclusions because they presume that the comparison of incompatible theories from adjacent explanatory levels (in this case, the psychological and the neuroscientific) at a particular point in the history of scientific inquiry will yield the same results that the comparison over time of incompatible theories from the *same* science yields. In short, presuming a single model of reduction, they apply the lessons that arise from *intra*level relations to a case of *inter*level relations.

Not coincidentally, one of the first eliminative materialists was Paul Feyerabend (1967). Feyerabend (1962) and Thomas Kuhn (1962) famously studied theory change over time within particular sciences and, thereby, questioned the logical empiricists' reductive model of progress within a science. The logical empiricists held that scientific progress turned on the discovery of ever more encompassing theories that not only predict, disclose, and organize previously unknown patterns of phenomena but also provide improved accounts of the phenomena that their predecessors explained. The new account is superior, if for no other reason than its ability to unify within the framework of a single set of theoretical principles what had previously appeared to be different domains. A well-worn illustration is the ability of Maxwell's theory to manage electrical, magnetic, and optical phenomena at once. Scientists subsequently regard the achievements of diverse earlier theories as special applications of the new, more general theory with which they are now armed. So, Galileo's law of free fall approximates an application of the principles of classical mechanics to falling objects comparatively close to the surface of the earth.

The logical empiricists alleged that the principles of the older theories follow as logical consequences from the principles of the new theory supplemented by a bit of logical machinery dealing with the translation of the theories' predicates. Feyerabend and Kuhn, however, argued on the basis of scrutinizing major episodes in the history of modern physical science that that supplementary translation machinery is overwhelmingly philosophical fiction, that the predicates of succeeding theories do not usually square in any straightforward way, if they square at all, with the predicates of their predecessors, and that change in the sciences frequently involves abrupt, discontinuous shifts in theory and practice. Many, if not most, of science's great theoretical successes do not fit neatly with the view of scientific progress as the accumulation of truths.

His criticisms of the logical empiricists' account of theory reduction in science notwithstanding, Feyerabend (at least at this point in his career) did not abandon their ideal of providing a single, unified model of the diverse matters that the logical empiricists lumped under the rubric of "reduction." Because he construed the progress marked by the succession of theories in particular sciences over time according to the same principles as the progress scientists achieve when they elaborate and enrich their models by looking to research going on at the same time in sciences operating at different analytical levels, Feyerabend arrived at eliminative materialism, i.e., he sketched a case for the liquidation of psychology and its accompanying ontology in deference to superior explanations at the neuroscientific level.

Clifford Hooker (1981), Paul and Patricia Churchland (1990), and John Bickle (1998) endorse what Bickle calls "New Wave" reductionism. New Wavers *explicitly* propose models of intertheoretic reduction in science that simultaneously aspire to incorporate Feyerabend and Kuhn's insights and to retain the logical empiricists' ideal of a single, multipurpose model. (For example, an *Ur*-text of this movement, Hooker, 1981), is entitled "Towards a General Theory of Reduction.") New Wave reductionists hold that the myriad cases of intertheoretic relations in science scatter across a continuum from better to worse intertheoretic connections. At one end of that continuum, where one theory or, at least, a theory-analogue, maps well onto another theory, these intertheoretic connections approach the ideal of a classical intertheoretic reduction of the sort the logical empiricists touted (Schaffner, 1967). At the opposite end of that continuum fall the kinds of cases that Feyerabend and Kuhn stressed. There the failures of intertheoretic mapping are sufficiently numerous and severe that any hope of mapping one theory onto another is forlorn. Feyerabend and Kuhn, as well as the New Wave reductionists they have inspired, noted that, without prospects for reconciling the two theories, scientists simply abandon the less accomplished theory and its distinctive ontology. (Consequently and, again, not coincidentally, one of the second-generation eliminative materialists was Paul Churchland (1981).) The history of modern science is a history of discarding cherished commitments about what the universe contains. The crystalline spheres, the bodily humors, and the phrenological faculties are all instances of the theoretical flotsam and jetsam that bobs aimlessly in the wake of scientific progress. When old theories cannot be readily mapped onto their superior successors, science *eliminates* both them and all of the stuff that only they discuss.

I (McCauley, 1986, 1996, in press) have argued that the elimination of psychology and its objects of study in favor of neuroscience and its objects of study that New Wave models envision and that eliminative materialists extol is implausible. I have pressed both normative and factual arguments. The latter is a simple historical induction. What both the eliminative materialists and the New Wave reductionists have continually ignored is that the dire outcomes they highlight do *not* arise in every circumstance where two theories with overlapping explananda conflict. *All* of the theory eliminations that Kuhn (especially) made famous and that the eliminative materialists seized upon to model the future of scientific explanations of human behavior occur in Wimsatt's *intralevel* settings, i.e., they concern changes over time *within* a particular science. The problem, however, is that the contemporary and future relations between research in neuroscience and psychology about which the eliminative materialists conjecture involve *interlevel*, i.e., cross-scientific, relations.

The (negative) normative point is that the kinds of interlevel inquiries that the psychology–neuroscience case instantiates are not abetted by the elimination of higher level psychological theories, let alone the entire elimination of the psychological sciences from which they spring, which the eliminativists have envisioned in some of their most extreme proposals (e.g., Churchland, 1979, 1981). This would simply deprive neuroscience of the theoretical and conceptual direction, the experimental techniques, and, most notably, the large bodies of evidence that the psychological sciences provide, to say nothing of largely marooning inquiries at the sociocultural level from substantial interlevel influences. Following the recommendations of the eliminative materialists and of the application of the New Wave model of reduction to this cross-scientific case would yield a science overall (and a neuroscience!) that possesses far more limited theoretical, practical, and evidential resources, which is to say that it would yield a science overall and a neuroscience, in particular, that are comparatively impoverished. Note, I am *not* saying either that eliminations in science do not occur or that they should not occur but, rather, that with established sciences, such as neuroscience and psychology, they do not occur as the direct result of cross-scientific conflicts of theory.

Wimsatt (1976), Patricia Churchland (1986), and Bickle (1998) have all adopted the language of the "co-evolution of theories" to describe (among other things, in Churchland and Bickle's cases) the ongoing negotiations of concepts, theories, and evidence that occur between scientific inquiries carried out at different levels of explanation in science. The central insight is that eventually scientific research projects carried out at adjacent levels of explanation will inevitably exert selection pressures on one another. All else being equal, scientists prefer theories that are consilient to theories that are not (Wilson, 1998).

The mechanists' analyses of specific explanatory problems, such as Bechtel's account (2006) of the study of cell mechanisms and of the emergence of cell biology in the middle of the twentieth century, supply the kind of detailed examinations of cases that reveal the limitations of the co-evolutionary talk for making sense of interlevel, i.e., multilevel, inquiries. Ultimately, the co-evolutionary metaphor is unsatisfactory, since one of the possible outcomes of the co-evolution of species is the extinction of one or the other or both. But as the mechanists' discussions show, the inquiries in question reliably result in the ever more complex integration of concepts, theories, practices, and evidence between what are, nonetheless, distinguishable analytical levels (see, for example, Craver's (2001) general model of mechanistic analysis). *None* of the historical cases of eliminations in science, i.e., the (theoretical) analogs of extinction in the history of science, that either the eliminativists or the New Wave reductionists cite are cases of interlevel relations, and this pronouncement applies even to the putative counter-examples that the Churchlands (1996) have offered (see McCauley, in press). Once *scientific disciplines* (and subdisciplines such as cell biology) achieve some stability, as marked not only by theoretical and empirical accomplishments but also by social developments (such as the emergence of distinctive disciplinary names and corresponding societies, journals, university departments, etc.), they, in contrast to particular theories that might rule in those disciplines at any specific moment, do not, subsequently, go extinct. Instead, they and their currently prevailing theories add to the ever more richly woven fabric of explanations and accounts of the world that the sciences furnish. They contribute to the variety of analytical and explanatory perspectives that explanatory pluralism emphasizes.

Two qualifications are in order before proceeding in the next sections to a new set of considerations that will only complicate this picture more. First, none of this is intended to slight the fundamental contribution of reductionist research strategies in science. Arguably, no discovery strategy has proven any more profitable than reductionism in the history of science (rather, the issues concern getting clear about just what various types of reductions do and do not amount to). Commentators have, if anything, faulted explanatory pluralists for their *optimism* about the promise of reductionist research strategies (Schouten & Looren de Jong, 2001).

Second, some subsequent writings on matters of the mind indicate that the Churchlands (e.g., P. M. Churchland, 1986; P. S. Churchland, 1996) have been inclined, at least some of the time, to relax their eliminativism, typically in favor of some version of the psycho-physical identity theory. Neither their subsequent writings on intertheoretic reduction (e.g., Churchland & Churchland, 1996) nor the subsequent writings of other New Wave reductionists (e.g., Bickle, 2003), however, budge in their support for a single, multipurpose model of intertheoretic reduction. The Churchlands offer some unconvincing counter-examples to the historical argument (McCauley, in press), which Bickle's discussions simply ignore. But, crucially, neither take up the normative arguments at all. This is unfortunate. For if the analyses I advance below are on the right track, then all of the philosophers who have written on cross-scientific relations and especially the New Wave reductionists, who have propounded a general, all-purpose model that presumes to supply exhaustive accounts of intertheoretic relations, of cross-scientific relations, and of scientific progress, have significantly *underestimated* the diversity of patterns that arise in scientific research.

3. A Further Reason why Reduction in Science Is Not Simple: Distinguishing Between Synchronic and Diachronic Modes of Scientific Theorizing

The kind of cross-scientific elimination of theories and of whole sciences that the New Wave reductionists dream about and the kind of cross-scientific (heterogeneous) reductions of *sciences* that Nagel (1961) explicitly discussed are unpromising for an additional reason.

To see why will turn on deploying a common enough distinction. It will be useful in thinking about cross-scientific relations to differentiate between *synchronic* scientific theories and research devoted to explaining *structural* phenomena as opposed to theorizing about *diachronic* matters devoted to explaining *processes, especially those processes that take place over comparatively long time frames on a comparatively large scale.* This is a distinction between two different modes of scientific theorizing and projects of research to which philosophers writing on reduction in science have given little attention.

That said, I should begin by noting some necessary caveats. This is not to say that diachronic theories have received *no* attention from philosophers interested in cross-scientific relations. For reasons connected with paths mechanistic analyses inevitably take, recent mechanistic discussions have earned the clearest exemption from such a charge. See, for example, Bechtel's discussion (2006, pp. 94–117) of the vital contributions to scientists' understanding of cell mechanisms in the first four decades of the twentieth century that biochemical research on the many *processes* involved in aerobic cellular respiration furnished. As that discussion illustrates, the fine-grained specification of mechanisms requires careful attention to the processes in which those mechanisms are involved. Still, both here and elsewhere in the literature on mechanistic analysis, any discussions of diachronic theorizing that spring up focus overwhelmingly on accounts of short term processes concerning small scale systems. I will return to these issues below.

Analyses of systems' *structures* (including analyzing the structures of mechanisms) concentrate on *compositional* relations. Such analyses decompose systems into their parts. Understanding the operations of systems and mechanisms partially depends upon tracing the spatial relations and the connections among their parts. This is, perhaps, the most transparent illustration of the reductionist impulse in scientific research available, and the philosophical literature on reduction in science is replete with examples. Up to now philosophical discussions of cross-scientific relations have mostly proffered such structural analyses and have, thereby, reinforced most philosophers' conceptions of these matters.

The emphasis in philosophical discussions of reduction on structural theories and compositional relations has been an important impetus for the widespread predilection to use mereological criteria to identify explanatory levels in science (Kim, 1998). The basic intuition is that big things are made of not-so-big things, that those not-so-big things are made of smaller things, and that those smaller things are made of things smaller yet. This is why philosophers so often rely on considerations of scale as a heuristic for distinguishing explanatory levels in science. This leads to what are often far more detailed conceptions of the hierarchical arrangement among the sciences (e.g., Churchland & Sejnowski, 1992, p. 11) than the broad categories employed at the left side of Figure 9.1.

Rooting accounts of explanatory levels in science in a mereological conception of organizational levels in nature, however, generates anomalies. For the physical sciences not only address the smallest and most fundamental building blocks of nature. They are simultaneously concerned with the very *largest* things as well. The basic physical sciences include both the subatomic and the astrophysical. So, a scheme for organizing the sciences that looks to mereological criteria inspired by reductionists' preoccupations with synchronic, structural theories results in descriptions that miss at least some of those sciences' projects. (In fact, a strategy for differentiating organizational levels in nature and corresponding explanatory levels in science that relies on *diachronic* considerations does a better job. At least, it readily accounts for the broad hierarchy of the families of the sciences portrayed in Figure 9.1. In short, the operative principle is that the longer the systems some science specializes in have been around, the lower that science's analytical level (McCauley, in press).)

The reductionist presumption is that the structures, patterns, and principles (that those structures and patterns inspire) at lower levels will suffice to explain the structures and patterns at higher levels and the principles they inspire. Logical empiricists aim to deduce the *principles*, i.e., scientific laws, at higher levels from the principles at lower levels, in combination, of course, with a translation apparatus (mentioned earlier) connecting, as systematically as possible, the predicates the laws include. So, for example, with the aid of appropriate translation machinery, the logical empiricist aspires to reduce the laws of chemical bonding to the laws in subatomic physics concerned with the actions of atoms' components. In cross-scientific settings the logical empiricist model seeks to reduce *theories and their laws* by explaining them in terms of the theories and their laws at lower levels of explanation.

As stressed earlier, critics, commentators (Schaffner, 1967), and New Wave reductionists all despaired of ever finding translation machinery sufficient to carry through classical theory reductions. Instead of mapping higher-level theories and their principles on to lower-level ones, the New Wave reductionists either speak of mapping theory analogs in the hopeful cases or of simply ignoring (if not dispensing with) the higher-level enterprises when the explanatory perspectives at higher and lower levels substantially diverge. Although one of my secondary agendas is to discredit these latter proposals, it is the former one about the hopeful cases which concerns me here.

Paul Churchland (1989, p. 49) holds that New Wave theory reduction involves the construction of an "equipotent image" of the higher-level, reduced theory within the framework of the laws and principles that the lower-level theory provides. But talk of "images" in the context of discussions of *logical* reconstructions of scientific theories and laws is metaphoric. Laws and theories are not the sorts of things that most people entertain *images* of. The emergence and prominence of the metaphor, however, is not coincidental. Careful examination of New Wave work on "theory reduction" over the past couple of decades reveals that rather than reconstructing higher-level *theories and laws* at lower levels, New Wave reductionists have often themselves been constructing equipotent images of upper-level *patterns and mechanisms* at lower levels. (Examples include Churchland, 1986, ch. 10; Churchland, 1989, pp. 77–110, 1995; Churchland & Sejnowski, 1992; Bickle, 2003.)

That New Wave reductionists have increasingly turned to tracing (equipotent) images at lower levels of patterns and mechanisms at higher levels certainly comes as no surprise to the mechanists, whose work, in effect, underscores the fact that the traditional and New Wave focus on scientific laws and theories requires a forced fit, at best, in the biological, psychological, and (I would add) sociocultural sciences (Wright & Bechtel, in press). These sciences typically traffic in phenomenal patterns and functionally characterized mechanisms (more than in laws) and in increasingly fine-grained models of those mechanisms (more than in broad, general theories). This is not to say that these sciences never involve general explanatory principles or ambitious, sweeping theories (e.g., Llinás, 2001) but rather to indicate that they are the exceptions rather than the rule. Robert Cummins (2000) observes, for example, that most of psychology's principles are not explanatory laws so much as *effects*, i.e., patterns at the psychological level in need of (further) explanation. Reductionists' interests in interlevel relations within and between the higher level families of the sciences, i.e., at levels of explanation higher than the basic physical sciences, mostly concern tracking the *structures and operations of mechanisms* as explanations of higher-level patterns rather than tracking the *logical connections (or the lack thereof) between theories and laws.*

Oddly, nothing about the uncontroversial metaphysical principles concerning compositional relations in synchronic, structural studies mandates or even especially favors reductionism. That such an orientation remains as popular as it does is, presumably, a function of its historical success in abetting scientific research. Scientists can, however, just as readily ask questions about the structure of the larger system or pattern to which the targeted part contributes (Craver, 2001). They can not only examine the context in order to ascertain the part's spatial relations and connections to other components, they can also examine the roles it plays in the characteristic processes the larger system exhibits. When researchers have reason to suspect that such arrangements are the results of *selection* (natural or otherwise), looking to higher levels can offer clues about systems' functions and, thereby, suggest criteria for individuating mechanisms on functional grounds as well as on spatial and structural ones.

If they do not perennially operate in the synchronic mode, structural theories, at least, start that way. They look at a system's parts without reference to time. Upon pondering a structure's function, however, whether that of the overall system or of one of its parts, the spotlight can shift to temporal considerations. At each level scientists search for interacting systems that exhibit coincident structural and operational patterns. Machinery in nature can display myriad unanticipated complexities (Bechtel, 2006). The first step back from the machinery's structural intricacies reliably spawns reflection on that machinery's organization, on its functions, and on the processes that contribute to realizing those functions.

Especially in the higher-level families of the sciences, where selective forces impinge and, thus, where functional considerations seem to possess greater analytical promise, structural investigations ineluctably prompt inquiries about functions, which in turn inspire investigations into processes and operations, i.e., into *changes* in these systems over time. This is why emphasizing the distinction between synchronic and diachronic phenomena and especially a corresponding distinction between synchronic and diachronic modes of analysis will probably feel somewhat idealized (to those who are sympathetic) or somewhat contrived (to those who are less so). What adds to those impressions is the fact that learning more about a system's structure provides clues about its function, and learning more about the accompanying processes can highlight structural details that might have otherwise gone unnoticed. These two sorts of analysis not only provide resources for one another's improvement but often are intimately intertwined. At the finest resolutions in the biological, psychological, and sociocultural sciences, where considerations of function serve as the fulcrum on which this distinction between these two modes of scientific analysis balances, whether structural, synchronic analysis of a system, on the one side, or diachronic analysis of the short-term processes it exhibits, on the other side, dominate in any explanation can shift back and forth from one problem to the next.

Still, structural, synchronic analyses often leave diachronic questions mostly unaddressed, and when they do take up such matters, the time frames they contemplate, at least *initially*, are of *comparatively short duration*. This is as true about ancient scientists' meditations on the structure of the cosmos (as revealed by celestial bodies' apparent motions) as it is about modern scientists' meditations on the mechanisms of mind as revealed by features of human brains and human performance. Ancient astronomers logged the consequences of the earth's daily and annual motions in the apparent movements of objects in the heavens for nearly two millennia before Hipparchus in 129 BCE recognized the precession of the equinoxes, the effect of a third wobbling motion of the earth around its axis, which takes nearly 26,000 years to complete a single cycle. Psychologists, neuroscientists, and cognitive scientists, until quite recently, spent more than a century overwhelmingly preoccupied with structural models of the mind/brain and with evidence concerning its changes and operations over very short durations that rarely even extended to the length of the normal human life span. (Often the study of any processes that structural analyses occasion are also of comparatively narrow *scope*. Patricia Churchland (1986), for example, opens the tenth chapter of her landmark book, *Neurophilosophy*, with a discussion of the relative dearth of theorizing in neuroscience and of the exceedingly fine-grained focus of most work in that field.)

Of course, with respect to scientific theorizing and research on *patterns of great duration* in the biological, psychological, and sociocultural sciences, Charles Darwin (1859/1979) is *the* pivotal figure. Darwin's greatness arises not only because he furnished modern science with a process, natural selection, that explained why biological systems can be profitably understood as carrying out functions without necessarily engendering worries about illicit teleological ascriptions. His theory of evolution by natural selection also *substantially expanded the time frames* that scientists consider when they theorize about the forces that have shaped not just the biological world but also the human mind.

Addressing long trajectories across natural history, Darwin's theory also requires *a new unit of analysis.* That analytical unit, a *species*, is extended in time and includes countless individual organisms. Darwin's theory of evolution infused the notion of a species with a *theoretical-explanatory* salience that it had never possessed before. Its newfound salience for theorizing on the diachronic front also far exceeded any interest the concept might have attained on the structural front theretofore. (In a biology of impregnable boundaries between species, large-scale structural relations are little more than curiosities, and they certainly offer no clues about patterns of descent.) The salience of species for diachronic theorizing has also served as a prominent wellspring for whatever theoretical interest this notion possesses on the structural front since. The subsequent synthesis of Darwinian evolution and modern genetics supplies detailed theoretical grounds for why recurrent structural features in a species' members can now be said to offer clues about what was once presumed to be an underlying "nature" that they share but is now explicated in terms of the similarities of their genomes.

The new emphasis on this large-scale unit of analysis (species) does not debar interest in the smaller-scale systems (individual organisms) that make it up. *Both* are perfectly appropriate objects for diachronic theorizing. The distinction in biology between ontogeny and phylogeny has, during the past few decades, served as a template for diachronic theorizing in the psychological sciences. The analogue of ontogeny is individual psychological development. As with the other sciences, diachronic interests first erupted into full-blown theorizing in scientific psychology with the study of relatively shorter time frames. Developmental psychology studies individual humans and, for example, their cognitive functioning across comparatively short time spans, typically no more than childhood, though subdisciplines concerned with adolescents, young adults, and the elderly have also arisen (presumably, such inquiries, in principle, examine no more than the full human life span).

In psychology, as with most of the other sciences, it has taken longer for systematic diachronic theorizing about the forces that impinge over immense time spans to emerge. The phylogeny of the human species is the analog at the level of biological theorizing for the evolution of the distinctively human mind at the psychological level. Here the interest is in those demands that, over great expanses of time, have shaped the structure and functioning of the minds of the ancestors of contemporary human beings. In the past two decades theories about what has come to be known as evolutionary psychology have emerged as the analogs of phylogenetic proposals in biology (e.g., Tooby & Cosmides, 1989, 1992). Evolutionary psychologists submit hypotheses about likely selection pressures that would have shaped the human mind for the longer term and conjectures about their expected implications for the *structure* of the modern human mind. Here too theories about long-term (large-scale) diachronic patterns impart newfound significance to debates about the mind's structure. The suggestion is that careful consideration and probing of, among other things, contemporary human behavior and mental life should yield evidence for these conjectures. Evolutionary psychologists contend that the resulting recurrent structures in the minds of individual *Homo sapiens* provide clues about underlying similarities of those individuals' genomes just as recurring structures in their bodily organs do. (This need entail neither genetic determinism nor a detailed genetic blueprint for the mind, though it does envision a comparatively fixed cognitive architecture that results from a characteristic developmental sequence across a broad range of divergent circumstances.) Such patterns in human mental life help identify plausible candidates for what might be broadly called "natural" features of the human mind. Such proposals have inspired dozens of new programs of experimental research (Buss, 1999).

In the face of their contentions about the evolution of the mind, of their comments on the unity of science, and of their insistence that no inquiry is "autonomous," that John Tooby and Leda Cosmides sometimes prove reluctant to acknowledge the importance of evidence about the structure and functioning of human *brains* for these matters is both puzzling and unfortunate (1992, pp. 19–24 and 65–66, respectively). Even the most preliminary exploration of the role of research on and theorizing about such diachronic matters will suggest that there are more cross-scientific influences than have generally been dreamt of in the philosophies of many philosophical champions of scientific reduction and, perhaps, of Tooby and Cosmides as well.

4. Complicating Reduction

Distinguishing between synchronic and diachronic modes of analysis in the various sciences and further distinguishing, among the diachronic analyses, between short-term processes in small-scale systems, such as the exceedingly brief influence of the visual icon on cognitive processing (Sperling, 1960) and extremely long-term processes in large-scale systems, such as the evolution of maturational proclivities in members of our species to acquire natural language (Pinker, 1994), complicates accounts of (1) analytical levels in science, (2) cross-scientific relations, and especially (3) theory reduction. Following are brief comments on each, in turn.

Introducing these distinctions suggests a more refined vision of analytical levels in science. Attention to these distinctions requires us to add a third dimension to the picture of analytical levels presented in Figure 9.1. Figure 9.2 retains Figure 9.1's resources to represent both intralevel relations (in the arrow of time across the bottom) and the interlevel relations between synchronic, structural theories (in the front plane for each of the families of the sciences). Figure 9.2, however, also depicts a third dimension, which permits the representation of theorizing and research in the diachronic mode. Research in the diachronic mode addresses two different time frames (at least) and, correspondingly, systems at two different scales. Thus, Figure 9.2 contains *two* additional planes at each level of analysis. They represent the distinction between theories about the workings of small-scale mechanisms in short-term time frames and accounts of change in large-scale systems over much longer periods of time. The first sort of diachronic project, for example, research in developmental psychology on children's growing command of theory of mind (Wellman, 1990), or research in neuroscience on long-term potentiation (Lynch, 2000), is represented in Figure 9.2 for each of the families of the sciences by the planes at the back. The second sort of diachronic project, which in psychology has emerged as a *bona fide* subdiscipline since

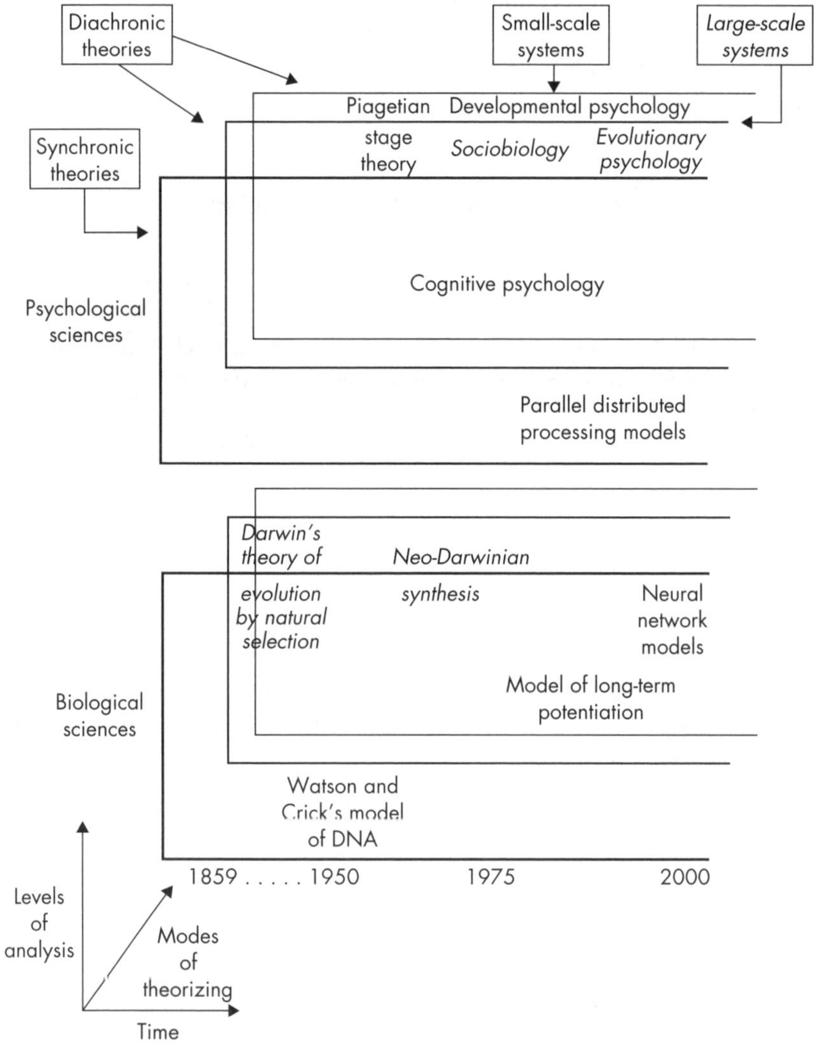

Figure 9.2 Introducing two sorts of diachronic theorizing

the mid-1990s, is represented in Figure 9.2 for each of the families of the sciences by the plane that is in between the other two. (The specific theories, models, and subdisciplines situated in Figure 9.2 are a disparate lot. The aim is only to illustrate the kinds of theories and research that the three planes represent within each family of sciences and the points in time that they arose.)

The most obvious implication of this more complex characterization of the varieties of theorizing and research that go on at each analytical level in science is that it *increases* the number of locales from which intertheoretic and cross-scientific influences can arise. Again, if your reflexive response is to suspect that this rather idealized account misses how much considerations of *function* in systems shape our understanding of their *structures* (and vice versa), or how, for example, our growing knowledge of the molecular structures of the mechanisms of heredity has influenced our understanding of ontogeny (and vice versa), fear not! For *highlighting such bidirectional influences between diachronic and synchronic projects is just the point*! What this more intricate picture of scientific endeavors at each analytical level makes clear is just how many more loci are available for cross-scientific influence, evidence, and inspiration, and examples of all of these possibilities abound. Perhaps most prominently of all, Darwin's theory of biological evolution on the basis of natural selection has served as the model for most subsequent theorizing about large scale change over the long term in both the psychological and the sociocultural sciences (e.g., Boyd & Richerson, 2005).

Limitations of space permit only a cursory review of some connected examples within cognitive science. Among Noam Chomsky's early proposals (1965, 1972, 1975) were the claims that linguistics is best conceived as a subdiscipline of psychology and that human beings are innately endowed with a task-specific language-acquisition device, i.e., a system of principles that constrains the forms of possible natural languages. In effect, Chomsky advanced a theory about some of the human mind's standard equipment.

These proposals invited psychologists of language to undertake psycholinguistic experimentation to test them. Those tests of what is, at its core, a structural theory often concerned the theory's implications for linguistic processing, i.e., for diachronic patterns concerned with local operations such as an application of a passive transformation. Ironically, Chomsky did not accord vast attention to the (mixed) results of this work (Abrahamsen, 1987; McCauley, 1987; Reber, 1987). So, Chomsky's many pronouncements about the place of linguistics within psychology notwithstanding, it is linguists and psychologists who are basically unsympathetic to his theories that have most thoroughly explored the connections between research in these two disciplines (Lakoff, 1987; Langacker, 1987; Gibbs, 2006).

However inert the connections between Chomsky's theories and most research in psycholinguistics has proved, his unwavering nativism concerning the language acquisition device and Jerry Fodor's (1983) related proposal about innate modules that constitute the mind's input systems were among the influences that inspired new *diachronic* theorizing within psychology (not all of that theorizing is sympathetic to their views, e.g., Deacon, 1997). For the past two decades evolutionary psychologists have formulated diachronic proposals about the selection pressures that might be responsible for recurrent features of the minds of contemporary human beings. Although these proposals have proven both suggestive and controversial, I wish here to stress but a single point. These diachronic speculations have, in turn, inspired new hypotheses about the mind's *structure*. Evolutionary psychologists have proposed that the mind is not merely modular at its periphery, as Fodor suggests, but is, instead, *massively* modular (Barkow, Cosmides, & Tooby, 1992; Pinker, 1997; Plotkin, 1998). A survey of the various inventories that Tooby and Cosmides (1992) supply suggests that the modern human mind possesses scores of domain specific capacities, at the very least. They readily appeal to research in experimental psychology concerned with the structure and functioning of various domain-specific systems that their large-scale, long-term diachronic theory predicts, for example, systems governing disgust reactions and the avoidance of contaminants (Rozin, Haidt, & McCauley, 2000).

Chomsky and Fodor's resuscitation of nativist theories in the cognitive sciences and the evolutionary psychologists' even bolder conjectures about such matters have come in an era when new bodies of evidence have emerged in sciences as diverse as molecular genetics, primatology, and experimental psychology that enable researchers to assess empirically those theories' commitments, both structural and diachronic. For example, on the basis of a review of the available *archaeological* evidence, Steven Mithen (1996) argues not only for a different account of the quasi-modular architecture of the contemporary human mind but an alternative view of its evolution over the past 60,000 years. In short, Mithen argues that Tooby and Cosmides' Swiss Army knife conception of the mind does, indeed, characterize the mind of *Homo sapiens* during much of the Pleistocene but that archaeological investigation supplies compelling evidence for subsequent evolutionary developments in human cognitive capacities during the last sixty or so millennia, resulting in a mind possessing what Mithen calls "cognitive fluidity" between otherwise largely modularized capacities.

Even this speedy overview of but a tiny sliver of the speculations and research during the past few decades in the psychological and cognitive sciences reveals intertheoretic influences that go back and forth between the three types of scientific projects that the planes in Figure 9.2 represent. But not only do diachronic projects influence structural ones within the same family of sciences, diachronic theories and research also exhibit cross-scientific influences *between* the families of sciences. Consider, for example, the fruitful interactions between psychological research on cognitive development and research on neural development (Munakata, Casey, & Diamond, 2004).

What impact should these considerations have on rethinking reduction? I end with three observations. First, even traditional accounts of reduction that pertain to structural matters suggest at least one plausible analogy for thinking about cross-scientific relations between theories and research in the diachronic mode. The "reductive" strategy here would be to show how our understanding of change at a higher level can be amplified and enhanced by looking both at changes at lower levels and at our best models of them. Analogy, however, is not identity. As the battery of statistical tools that have been developed for such purposes indicate, differentiating and analyzing the relationships of *component processes* looks like a considerably more difficult and complicated task than differentiating and analyzing the relationships of component structures and mechanisms.

Second, because cross-scientific relations often involve *evidential* connections bearing on scientific justification, reductive proposals suggesting that these connections are merely heuristic or peripheral or inconsequential from the standpoint of epistemology miss the mark, even on conservative accounts of scientific progress. Since nearly all of the participants in these discussions are *naturalists* in the philosophy of science, it is all the more perplexing that their proposals would also exclude or downplay the contributions of theories and research about diachronic matters in psychology. Consider, for example, the provocative implications for future research in neuroscience of the discovery of developmental prosopagnosia (e.g., Barton, Cherkasova, & O'Connor, 2001). In such a light the future science of human brains, behavior, and mental life that some reductionists in philosophy envision looks all the more impoverished.

Finally, pondering the place of scientific work in the diachronic mode only reinforces the pragmatic morals of explanatory pluralism. Exploring cross-scientific connections of all sorts is a valuable strategy of discovery at every explanatory level in science that enables researchers to find and exploit new theoretical and evidential resources. Reduction is a priceless tactic in science, but that is not the same thing as a metaphysical program and it engenders no reasons to anticipate the dispensability, let alone the eradication, of sciences.

The additional intertheoretic and cross-scientific relations that distinguishing between science in its diachronic and synchronic modes introduces augur not for simpler models of reduction but for more complicated ones.

References

Abrahamsen, A. (1987). Bridging boundaries versus breaking boundaries: Psycholinguistics in perspective. *Synthese*, *72*, 355–388.

Barkow, J. H., Cosmides, L., & Tooby, J. (Eds.). (1992). *The Adapted Mind: Evolutionary Psychology and the Generation of Culture*. New York: Oxford University Press.

Barton, J., Cherkasova, M., & O'Connor, M. (2001). Covert recognition in acquired and developmental prosopagnosia. *Neurology*, *57*, 1161–1168.

Bechtel, W. (2006). *Discovering Cell Mechanisms: The Creation of Modern Cell Biology*. New York: Cambridge University Press.

Bechtel, W., & Richardson, R. C. (1993). *Discovering Complexity: Decomposition and Localization as Strategies in Scientific Research*. Princeton: Princeton University Press.

Bickle, J. (1998). *Psychoneural Reduction: The New Wave*. Cambridge, MA: MIT Press.

Bickle, J. (2003). *Philosophy and Neuroscience: A Ruthlessly Reductive Account*. Dordrecht: Kluwer Academic Publishers.

Boyd, R., & Richerson, P. J. (2005). *The Origin and Evolution of Cultures*. New York: Oxford University Press.

Buss, D. (1999). *Evolutionary Psychology: The New Science of the Mind*. Boston: Allyn & Bacon.

Chomsky, N. (1965). *Aspects of a Theory of Syntax*. Cambridge, MA: MIT Press.

Chomsky, N. (1972). *Language and Mind*. New York: Harcourt Brace Jovanovich.

Chomsky, N. (1975). *Reflections on Language*. New York: Pantheon.

Churchland, P. M. (1979). *Scientific Realism and the Plasticity of Mind*. Cambridge: Cambridge University Press.

Churchland, P. M. (1981). Eliminative materialism and the propositional attitudes. *Journal of Philosophy*, *78*, 67–90.

Churchland, P. M. (1986). Some reductive strategies in cognitive neurobiology. *Mind*, *95*, 279–309.

Churchland, P. M. (1989). *A Neurocomputational Perspective: The Nature of Mind and the Structure of Science*. Cambridge, MA: MIT Press.

Churchland, P. M. (1995). *The Engine of Reason, the Seat of the Soul: A Philosophical Journey into the Brain*. Cambridge, MA: MIT Press.

Churchland, P. M., & Churchland, P. S. (1990). Intertheoretic reduction: A neuroscientist's field guide. *Seminars in the Neurosciences*, *2*, 249–256.

Churchland, P. M., & Churchland, P. S. (1996). McCauley's demand for a co-level competitor. In R. McCauley (Ed.), *The Churchlands and Their Critics* (pp. 222–231). Oxford: Blackwell.

Churchland, P. S. (1986). *Neurophilosophy: Toward a Unified Science of the Mind–Brain*. Cambridge, MA: MIT Press.

Churchland, P. S. (1996). Do we propose to eliminate consciousness? In R. McCauley (Ed.), *The Churchlands and Their Critics* (pp. 297–300). Oxford: Blackwell.

Churchland, P. S., & Sejnowski, T. J. (1992). *The Computational Brain*. Cambridge, MA: MIT Press.

Craver, C. F. (2001). Role functions, mechanisms and hierarchy. *Philosophy of Science*, *68*, 53–74.

Cummins, R. (2000). "How does it work?" versus "What are the laws": Two conceptions of psychological explanation. In F. Keil & R. Wilson (Eds.), *Explanation and Cognition* (pp. 117–144). Cambridge, MA: MIT Press.

Darwin, C. (1859/1979). *The Origin of Species*. Harmondsworth: Penguin.

Deacon, T. W. (1997). *The Symbolic Species: The Co-Evolution of Language and the Brain*. New York: Norton.

Feyerabend, P. K. (1962). Explanation, reduction, and empiricism. In H. Feigl & G. Maxwell (Eds.), *Minnesota Studies in the Philosophy of Science* (Vol. 3, pp. 28–97). Minneapolis: University of Minnesota Press.

Feyerabend, P. K. (1967). Materialism and the mind–body problem. *The Review of Metaphysics*, *17*, 49–67.

Fodor, J. A. (1983). *The Modularity of Mind*. Cambridge, MA: MIT Press.

Gibbs, R. (2006). *Embodiment and Cognitive Science*. New York: Cambridge University Press.

Glennan, S. (1996). Mechanisms and the nature of causation. *Erkenntnis*, *40*, 50–71.

Hooker, C. (1981). Towards a general theory of reduction. *Dialogue*, *20*, 38–59, 201–236, 496–529.

Kim, J. (1998). *Mind in a Physical World: An Essay on the Mind–Body Problem and Mental Causation.* Cambridge, MA: MIT Press.

Kuhn, T. (1962). *The Structure of Scientific Revolutions.* Chicago: University of Chicago Press.

Lakoff, G. (1987). *Women, Fire, and Dangerous Things: What Categories Reveal about the Mind.* Chicago: University of Chicago Press.

Langacker, R. (1987). *Foundations of Cognitive Grammar.* Stanford: Stanford University Press.

Llinás, R. (2001). *I of the Vortex.* Cambridge, MA: MIT Press.

Lynch, G. (2000). Memory consolidation and long-term potentiation. In M. Gazzaniga (Ed.), *The New Cognitive Neurosciences* (pp. 139–157). Cambridge, MA: MIT Press.

Machamer, P., Darden, L., & Craver, C. F. (2000). Thinking about mechanisms. *Philosophy of Science*, *67*, 1–25.

McCauley, R. N. (1986). Intertheoretic relations and the future of psychology. *Philosophy of Science*, *53*, 179–199.

McCauley, R. N. (1987). The not so happy story of the marriage of linguistics and psychology or how linguistics has discouraged psychology's recent advances. *Synthese*, *72*, 341–353.

McCauley, R. N. (1996). Explanatory pluralism and the coevolution of theories in science. In R. N. McCauley (Ed.), *The Churchlands and Their Critics* (pp. 17–47). Oxford: Blackwell.

McCauley, R. N. (2000). The naturalness of religion and the unnaturalness of science. In F. Keil & R. Wilson (Eds.), *Explanation and Cognition* (pp. 61–85). Cambridge, MA: MIT Press.

McCauley, R. N. (in press). Reduction: Models of cross-scientific relations and their implications for the psychology–neuroscience interface. In P. Thagard (Ed.), *Handbook of the Philosophy of Psychology and Cognitive Science.* Amsterdam: Elsevier.

McCauley, R. N., & Bechtel, W. (2001). Explanatory pluralism and the heuristic identity theory. *Theory and Psychology*, *11*, 738–761.

Mithen, S. (1996). *The Prehistory of the Mind.* London: Thames & Hudson.

Munakata, Y., Casey, B. J., & Diamond, A. (2004). Developmental cognitive neuroscience: Progress and potential. *Trends in Cognitive Science*, *8*, 122–128.

Nagel, E. (1961). *The Structure of Science: Problems in the Logic of Scientific Explanation.* New York: Harcourt, Brace, and World.

Oppenheim, P., & Putnam, H. (1958). Unity of science as a working hypothesis. In H. Feigl, M. Scriven, & G. Maxwell (Eds.), *Minnesota Studies in the Philosophy of Science* (Vol. 2, pp. 3–36). Minneapolis: University of Minnesota Press.

Pinker, S. (1994). *The Language Instinct: How the Mind Creates Language.* New York: HarperCollins.

Pinker, S. (1997). *How the Mind Works.* New York: Norton.

Plotkin, H. (1998). *Evolution in Mind: An Introduction to Evolutionary Psychology.* Cambridge, MA: Harvard University Press.

Reber, A. (1987). The rise and (surprisingly rapid) fall of psycholinguistics. *Synthese*, *72*, 325–339.

Rozin, P., Haidt, J., & McCauley, C. R. (2000). Disgust. In M. Lewis & J. M. Haviland Jones (Eds.), *Handbook of Emotions* (2nd ed., pp. 637–653). New York: Guilford Press.

Schaffner, K. (1967). Approaches to reduction. *Philosophy of Science*, *34*, 137–147.

Schouten, M., & Looren de Jong, H. (2001). Pluralism and heuristic identification: Some explorations in behavioral genetics. *Theory and Psychology*, *11*, 796–807.

Sperling, G. (1960). The information available in a brief visual presentation. *Psychological Monographs*, *74*, 1–29.

Tooby, J., & Cosmides, L. (1989). Evolutionary psychology and the generation of culture, Part I: Theoretical considerations. *Ethnology and Sociobiology*, *10*, 29–49.

Tooby, J., & Cosmides, L. (1992). The psychological foundations of culture. In J. Barkow, L. Cosmides, & J. Tooby (Eds.), *The Adapted Mind* (pp. 19–136). New York: Oxford University Press.

Wellman, H. M. (1990). *The Child's Theory of Mind.* Cambridge, MA: MIT Press.

Wilson, E. O. (1998). *Consilience: The Unity of Knowledge.* New York: Vintage.

Wimsatt, W. C. (1976). Reductionism, levels of organization, and the mind–body problem. In G. Globus, G. Maxwell, & I. Savodnik (Eds.), *Consciousness and the Brain* (pp. 205–267). New York: Plenum.

Wright, C. D., & Bechtel, W. (2006). Mechanisms and psychological explanation. In P. Thagard (Ed.), *Handbook of Philosophy of Psychology and Cognitive Science* (pp. 31–79). New York: Elsevier Science.

PART III

MECHANISMS OF MIND

10

COUPLING, EMERGENCE, AND EXPLANATION

Andy Clark

1. Introduction: Emergence and Explanation

"Artificial Life" names a very broad church indeed. The banner encompasses work on colonies of simple simulated organisms, all kinds of "synthetic biology," studies of real-world robots and autonomous agency, and much else besides. (For a representative sampling, see the variety of papers in volume 1 of the journal *Artificial Life*.) Underlying this surface variety, however, are a few relatively central and recurrent themes. Two such (related) themes are (1) the unfolding of patterns over time and (2) the emergence of interesting or adaptive features from clever couplings between agents and environments (including other agents).

This chapter focuses on these key themes as they appear in work on Autonomous Agents, Dynamic Systems Theory, and systems-level neuroscientific conjecture. Attention to this range of example cases can, I believe, help us to assess the more general role of explanations of emergent phenomena in cognitive science. Such a project seems worthwhile for at least two reasons. First, because an emphasis on emergent phenomena is highly characteristic of much of the most recent, challenging, and exciting work in Cognitive Science (from, for example, Mitchel Resnick's work on collective, decentralized behavior (Resnick, 1994a), to Luc Steels's work on behavior systems (Steels, 1994), to Harvey, Husbands, and Cliff's work on evolutionary robotics (Harvey et al., 1994)). But second, because some of the most basic questions about the explanation of emergent properties remain unresolved. Such questions include: how to define emergence itself; what general explanatory framework is best able to account for emergent features (is dynamic systems theory, for example, a better bet than classical component-based analysis?); and do such novel modes of explanation merely augment or radically displace more familiar styles of Cognitive Scientific understanding?

I shall argue that emergent phenomena do require new modes of explanation and understanding. But these modes do not displace more familiar projects such as homuncular decomposition and representational/computational description. Instead, we must use a variety of tools to understand the multiple aspects of real-time, embodied, embedded cognition. Two major explanatory projects are distinguished. One concerns the contribution of neurophysiologically real components to the psychologically characterized abilities of an agent. This project, I suggest, requires the use of some quite traditional analytic tools. And it yields a kind of understanding which is essential if we hope to explain, for example, the effects of local damage to system behavior. The other concerns the overall dynamics of the agent/environment system. This kind of understanding is essential if we hope to explain the adaptive success of embodied, embedded agents. The task of cognitive science, I argue, is to develop and carefully *interlock* both types of explanation.

2. Three Explanatory Styles

In this section I distinguish three styles of Cognitive Scientific explanation. The styles are quite general and cross-classify the use of particular programming styles such as connectionist vs. classicist.

2.1 Homuncular Explanation

To explain the functioning of a complex whole by detailing the individual roles and overall organization of its parts is to engage in homuncular explanation. This is the natural explanatory style to adopt when, for example, we explain the workings of a car, a television set, or a washing-machine. We explain the capacities of the overall system by adverting to the capacities and roles of its components, and the way they interrelate.

Homuncular explanation, thus construed, is the contemporary analog to good old-fashioned *reductionistic* explanation. I avoid the vocabulary of reduction for two reasons. First, much of the philosophical discussion about reduction assumed that reduction named a relation between theories, and that theories were linguaform, law-involving constructs. But in many cases (especially biology and artificial intelligence), what we might otherwise naturally think of as reductive explanations do not take this form. Instead, they involve the development of partial *models* which specify components and their modes of interaction, and explain some high-level phenomena (e.g., being a television receiver) by adverting to a description of lower-level components and interactions (see, e.g., Bechtel & Richardson, 1992). These are reductionist explanations in a broader sense – one which "homuncular explanation" seems to capture. And second, because to contrast emergent explanation with reductionist explanation would be to invite a common misunderstanding of the notion of emergence, that is to suggest that emergentist accounts do not explain how higher-level properties arise as a result of more basic structures and interactions. As we shall see later, recent emergentist hypotheses are by no means silent on such matters. Rather, the contrast lies in the *ways* in which the lower-level properties and features combine to yield the target phenomena. This kind of emergentist explanation is really a special case of reductionist explanation, at least as intuitively construed, since the explanations aim to render the presence of the higher-level properties unmysterious by reference to a multitude of lower-level organizational facts. (For an argument that emergence is best treated as a species of reduction, see Wimsatt (1986, 2000).) For these reasons, then, it will be more accurate and less confusing to contrast emergent explanation with homuncular explanation than with reductionist theorizing in general.

Modular programming methods in classical Artificial Intelligence (see, e.g., Newell & Simon, 1976; Haugeland, 1981) lent themselves quite nicely to a homuncular form of explanation. In attempting to understand the success of such a program, it is often fruitful to isolate the various sub-routines, etc., and to display their role in dividing up the target problem into a manageable series of sub-problems (see Dennett, 1978). The subsystems which such explanations identify coordinate their activity by passing messages, and are individually understood as operating upon symbolic input encodings so as to effect some desired transformation.

Much connectionist work, as Wheeler (1994) points out, is likewise apt for a kind of homuncular explanation. Solutions to complex problems (e.g., handwritten zip code recognition – Le Cun et al., 1989) exploit highly structured, multilayered networks (or networks of networks). In such cases it is possible to advance our understanding of how the system succeeds by asking after the roles of these gross components (layers or sub-nets). It is worth noting that homuncular explanation is most compelling when the components admit of straightforward representational interpretation, that is, cases in which the target systems have "reliably identifiable internal configurations of parts that can be usefully interpreted as representing aspects of the domain . . . and reliably identifiable internal components that can be usefully interpreted as algorithmically transforming those representations" (Beer, 1995, p. 225).

We shall return to this possible affinity between homuncular explanation and the positing of internal representations later on.

2.2 Interactive Explanation

This is less familiar and therefore merits a slightly more extended discussion. Unlike homuncular explanation, interactive explanation takes very seriously the role of the environment in promoting successful problem-solving activity. It seeks to display the ways in which crucial problem-solving moves may actively exploit the opportunities which the real world presents to embodied, mobile agents. (For a nice treatment, see Kirsh, 1995.) Such an explanatory strategy is not intrinsically in competition with the kind of homuncular approach reviewed above. It merely adds a new dimension of study and interest.

Work in the so-called "Animate Vision" (also called active vision, interactive vision) paradigm provides a clean example. At the heart of this paradigm (see Ballard, 1991; Churchland, Ramachandran, & Sejnowski, 1994) lies a challenge to the traditional construction of the problem of vision itself. The traditional construction (Marr, 1982) depicts the task of vision as the task of building a detailed representation of a 3-D world on the basis of what is essentially a body of 2-D data. From a certain perspective, the characterization seems obvious. Yet the problem of biological vision, Ballard suggests, is really quite different. In place of the task of 2-D to 3-D mapping he depicts the goal of vision as the production of successful actions within an environmental context, keeping computational costs as low as possible (see Ballard, 1991). Our conception of the very task of vision, it seems, may have been distorted by an atemporal, disembodied bias.

Once we confront the problems of vision on their home turf (the acting organism in a complex world), a new set of problems and resources comes to the fore. Thus we immediately note that the human visual system exploits multiple gaze-control mechanisms so as to get the most out of the very small area of high-resolution information provided by the fovea. We must constantly saccade to various locations in visual space and do so in response to a variety of cues. Thus imagine your task is to discover the 35-mm film display in a crowded drugstore. The canny cognizer, Ballard suggests, may here rely on a cheap cue detectable at the low-resolution peripheries of our vision: color. More precisely, she will move about seeking the low-level cue of Kodak yellow and, having detected it, be in a position to reorient her gaze so as to saccade in on the fine details of the film display (see Ballard, 1991, p. 71). Or suppose the task is to locate your car keys. The best strategy is to link the small object (the keys) in memory to a larger (more easily detectable) one such as a table. You can then seek the large object (the table) identifiable at low resolution and, having located it, saccade around until the keys are (hopefully) located. At this point we should note also the possibility for an active agent to alter the environment so as to increase the number of ways it may be exploited, for example, deliberately placing a colorful saucer on the table and always putting the car keys on it.

Interactive explanation, as remarked above, is fully compatible with a homuncular perspective on the agent's innards. Ballard posits, for example, a "stored model database" as an internal component whose role is to encode object-linking information. The contrast between the interactive approach and classical homuncular theorizing lies only in the way that close attention to environmental needs and opportunities is allowed to influence our conjectures about the components and their roles. Thus, according to Ballard, the idea of a component which encodes a full-scale model of our surroundings is misguided. Animate Vision, Ballard argues, neither needs nor can afford to create and sustain such a model. Instead, we constantly saccade around, picking up only such fragments of information as we need to support specific actions, and revisiting the scene again and again rather than relying on some internally represented surrogate.

2.3 Emergent Explanation

Emergent explanation is at once the most radical and most elusive member of our trinity. Where interactive explanation is usually just a sensitive and canny version of homuncular explanation, emergent explanation aims to offer a whole new perspective on adaptive success. At the heart of this new perspective lies the idea of organism–environment interactions which yield types of adaptive behavior not neatly attributable to any specific inner component or system. It is, however, no easy matter to give a more precise and clearly nontrivial account of this idea of emergence. Some writers associate emergence with unexpected/unprogrammed behaviors; but this yields an overly "observer-relative" notion of emergence. A better route, pursued by Steels (1994), is to restrict the use of emergence-talk to cases in which adaptive success is tied to the specifics of an ongoing interaction between the organism and the environment, *and* in which the patterns of results which this interaction yields require description in a vocabulary which differs from the one we use to characterize the powers and properties of the inner components themselves. Steels gives the example of emergent chemical properties such as temperature and pressure; these terms do not figure in our descriptions of the motion of individual molecules, but are needed to describe the behavior of aggregates of such items.

Steels's account introduces a useful distinction between controlled and uncontrolled variables. Controlled variables track behaviors or properties which can be directly affected by the organism. Uncontrolled variables track behaviors or properties which can be affected only indirectly, for example, by affecting some other gross action of the system. As Steels puts it:

> A controlled variable can be directly influenced by a system, for example, a robot can directly control its forward speed . . . An uncontrolled variable changes due to actions of the system, but the system cannot directly impact it, only through a side-effect of its actions. For example, a robot cannot directly impact its distance to a wall; it can only change its direction of movement which will then indirectly change the distance. (1994, p. 90)

This account applies nicely to classic examples of emergence such as Hofstadter's story about the operating system which begins to fail, to "thrash around," once about thirty-five users are online. In such a case, Hofstadter notes, it would be a mistake to go to the systems programmer and ask to have the "thrashing number" increased to, say, 60. The reason is that the number 35 is not determined by an inner variable upon which the programmer can directly act. Instead: "That number 35 emerges dynamically from a host of strategic decisions made by the designers of the operating system and the computer's hardware and so on. It is not available for twiddling" (Hofstadter, 1985, p. 642).

Here we have a fully *systems-internal* version of an uncontrolled variable. In this case affecting the variable would still require an indirect route, but not via environmental action but via adjustments to a host of inner subsystems whose collective behavior fixes the value of the variable. Emergent phenomena, as I shall use the term, are thus any phenomena whose roots involve uncontrolled variables and are thus the products of collective activity rather than of dedicated components or control systems. Emergent phenomena, thus understood, are neither rare nor breathtaking; nonetheless, getting target behaviors to arise as functions of uncontrolled variables has not been a common strategy in AI, and such behaviors, when they arise, demand types of understanding and explanation which go beyond both the homuncular and the interactive models rehearsed above.

Two simple examples will help. The first is drawn from Resnick (1994b) and concerns a strategy for getting simulated termites to collect wood chips and gather them into piles. One solution would be to program the termites to take chips to predesignated spots. Relative to such a solution, chip-piling would count as a controlled variable as piling behavior would be under direct control and would be fully "twiddle-able." An emergentist solution, by contrast, yields the behavior indirectly via the combined effects of two simple rules and a restricted environment. The rules were: "If you are not carrying anything and you bump into another wood chip, pick it up; If you are carrying a wood chip and you bump into another wood chip, put down the wood chip you are carrying" (Resnick, 1994b, p. 234).

It is not obvious that such a strategy will work, as it allows chips to be removed from piles as easily as they can be added! Nonetheless, 2,000 scattered chips, after 20,000 iterations, became organized into just 34 piles. The reason the piling behavior ends up overwhelming the de-piling behavior is that whenever (by chance) the *last* chip is removed from an incipient pile, that location is effectively blocked; no *new* pile can ever begin there, given the two rules. Over time, then, the number of possible pile locations in the artificial grid diminishes, forcing the chips to congregate in the remaining locations. It is the unprogrammed feature of "location blocking" which enables piling activity to outrun de-piling activity! In this example, it is clear that piling behavior is not directly controlled, but emerges as a function of the interplay between the simple rules and the restricted environment.

A second example is drawn from Hallam and Malcolm (1994). They describe a simple solution to the problem of getting a robot to follow walls. You equip the robot with an inbuilt bias to veer toward the right, and locate a sensor on its right-hand side which is activated by contact and which causes the device to turn a little to the left. Such a robot will, on encountering a wall on the right, first move away (thanks to the sensor) and then quickly veer back to reencounter the wall (thanks to the bias). The cycle will repeat and the robot will follow the wall by in effect repeatedly "bouncing off it." The point to notice is that the behavior of wall following here emerges as an uncontrolled variable out of the interaction between the robot and its environment. It is not subserved by any internal state encoding a goal of wall following. We, as external theorists, lay on the wall-following description as a gloss on the overall embedded behavior of the device. But the gloss does not attach to any *agent-side* inner states.

The distinction between controlled and uncontrolled variables is important but does not yet isolate what seem to me to be the most interesting cases of emergence. These are cases in which the uncontrolled variable tracks some process (which may or may not span brain, body, and world) involving "continuous reciprocal causation" (Clark, 1997a; 1997b). Continuous reciprocal causation, as there defined, is causation that involves multiple simultaneous interactions and complex dynamic feedback loops, such that (a) the causal contribution of each component in the system of interest is determined by, and helps to determine, the causal contributions of (often a large number of) other components, and (b) those contributions may, as a result, change quite radically as the process evolves. It is noteworthy that despite my use of the word "continuous," nothing in that definition requires real continuity in the signals exchanged. Cases such as these, I argued (Clark, 1997b; Wheeler and Clark, 1999), present the greatest challenge to the kinds of classical "homuncular" explanatory strategies of traditional cognitive science. The reason is that these cases, more than any others, seem to demand a shift to a larger unit of analysis.

To bring the idea into focus, consider the contrast between e-mail exchanges and face-to-face conversation. Fogel (1993) notes that an e-mail exchange conforms closely to the model of nonemergentist explanation described in the paper: in fact, it conforms to the very simple model that Clark (1997a) calls "catch and toss." Here, each agent affects the other, but they do so in a strictly turn-taking manner (this is true even if message sometimes cross in the ether). Now compare the case of a face-to face exchange. In the face-to-face condition, there is rich ongoing monitoring and adjustment, with input coming in very richly across many channels. There is a strong intuition (Fogel, 1993; see also Mascolo & Harkins, 1998) that such cases demand a move to a larger unit of understanding and explanation. The question is, WHY does this kind of signaling regime make a difference, and why should THIS kind of difference, in particular, be the kind that underwrites the move to a larger unit of analysis and explanation?

One possible answer (favored by Fogel – see also Ross, 2004) is that there is something special about continuous or analog signaling systems. But I now think that this may be subtly mistaken. What matters (I suggest) is not the presence of genuine continuity so much as the presence of ongoing exchange and adjustment over many channels. It is this complexity, rather than continuity *per se*, that makes room for very tiny seeds to take root and to initiate positive feedback cycles that rapidly lead to new outcomes. Here is a toy example:

> I give a very slight flicker of a smile as, in the course of a simple encounter, you say P. That encourages you to try out a gentle probe Q. I make a slight encouraging postural motion as you deliver Q, and I try out the slightly stronger R. You let your smile persist a fraction longer and offer S. I make a slightly more encouraging body movement at some critical moment during your delivery of S, and follow though with my own (now more or less undisguised) T.

Now the speakers are on the same page and (crucially) it is a page they would never have gotten to so rapidly had they been exchanging straight serial messages along a single and non-overlapping dimension.

Notice that nothing here depends on strict continuity (analogicity) in any of the signals. Two beings simultaneously deploying multiple fine-grained yet absolutely digital signaling systems could engage in just this kind of reciprocal dance and get all the benefits. Yet this, I currently feel, is the heart of it. What matters is not continuity *per se* but subtlety, richness, and overlap (so you can change course midway, alter emphasis according to ongoing feedback, etc.). What matters, in short, is just the complexity of the space in which positive feedback loops are able to take root. This leads immediately to a rather more graded understanding of emergence itself, since systems will be open to such seed-and-feedback-driven change to many different degrees depending on (a) the complexity and multiplicity of links and circuits, and (b) the temporal dynamics that govern their interactions.

Instead of resting with the simple distinction between controlled and uncontrolled variables, we thus note that the most important and challenging cases typically arise when the uncontrolled variable itself tracks a quantity (in this toy case, speed of convergence on a shared opinion) whose value depends upon the possibility of small seeds taking root within a complex, highly interconnected system. This leads to a cascade of episodes of reciprocal influence in which new features emerge and new solutions are found. Such cases, it seems to me, are the ones that most clearly warrant the shift to a larger unit of analysis. Such an analysis is, as we shall next see, merely one element in any full account, and will need to be supplemented by complementary analyses of the component elements.

3. Dynamic Systems and Emergent Explanation

What is the most effective explanatory framework for understanding emergent phenomena? A widely shared negative intuition is that classical homuncular explanation, at least, often fares badly in such cases (Steels, 1994; Maes, 1994; Wheeler, 1994). There are two rather distinct reasons for such failure.

One reason turns on the fact that many (not all) emergent cognitive phenomena are rooted in factors which spread across an organism and its environment. In such cases (and we saw several examples above) we ideally require an explanatory framework which is (1) well suited to modeling both organismic and environmental parameters and (2) models them both in a uniform vocabulary and framework, thus facilitating understanding of interactions between the two. A framework which invokes computationally characterized, information-processing homunculi is not, on the face of it, ideal as a means of satisfying (1) and (2).

A second reason turns on the nature of components. When the component parts each make a distinctive special contribution to the ability of a system to display some target property, homuncular analysis is a powerful tool. But some systems are highly homogenous at the component level, with most of the interesting properties dependent solely upon the aggregate effects of simple interactions amongst the parts. An example (noted by, e.g., van Gelder, 1991; Bechtel & Richardson, 1992) would be a simple connectionist network in which the processing units are all markedly similar and the interesting abilities are largely a function of the organization (by weighted, dense connectivity) of those component parts. A more complex case occurs when a system is highly nonhomogenous, yet the contributions of the parts are highly interdefined, that is, the role of a component C at time t_1, is determined by (and helps determine) the roles of the other components at t_1 and may even contribute quite differently at a time t_2, courtesy of the complex feedback and feed-forward links to other subsystems. Internal nonhomogeneity and online functional specialization are thus no guarantee that a homuncular analysis will constitute the most revealing description.

These complexities are reflected in Wimsatt's elegant description (1986) of "aggregate systems." Aggregate systems are the ones for which homuncular explanation is best suited. Such systems are defined as ones in which the parts could display their explanatorily relevant behavior even in isolation from one another, and in which the properties of a small number of subsystems can be invoked to explain interesting systemic phenomena (see also Bechtel & Richardson, 1992). As the complexities of interaction between parts increases, so the explanatory burden increasingly falls not on the parts but on their organization. At such times, we are driven to seek new kinds of explanatory frameworks. As we shall see later, it is likely that advanced biological cognition falls somewhere midway along this continuum. The systems have distinct and functionally specialized neural components. But the interactions (feedback and feed-forward relations) between these components are crucial determinants of most intuitively "psychological" phenomena. Good explanations, in such cases, require *both* a traditional homuncular explanation and something else besides. But what else?

Given our two desiderata (that we accommodate both organism/environment interactions and complex interactions between internal components), it is natural to consider the framework of Dynamical Systems Theory. Dynamical Systems Theory (Abraham & Shaw, 1992) provides a set of tools for describing the evolution of system states over time. In such descriptions, the theorist specifies a set of parameters whose collective evolution is governed by a set of (usually) differential equations. One key feature of such explanations, for our purposes, is that they are easily capable of spanning organism and environment. In such cases the two sources of variance (the organism and the environment) are treated as coupled dynamical systems whose mutual evolution is described by a specific set of interlocking equations. The behavior of a wall-mounted pendulum placed in the environmental setting of a second such pendulum provides an easy example. The behavior of a single pendulum can be described using simple equations and theoretical constructs such as attractors and limit cycles. (We need not unpack these terms here – but see, e.g., Norton (1995) for a good introduction.) But two pendulums placed in physical proximity tend, surprisingly, to become swing-synchronized over time. This synchronization admits of an elegant dynamic systems explanation which treats the two pendulums as a single coupled system in which the motion equation for each pendulum includes a term which represents the influence of the other's current state (the coupling in this case is achieved via vibrations running through the wall). For a fuller account see Saltzman (1995).

Dynamical Systems explanations are thus naturally suited to spanning multiple interacting components and whole agent-environment systems. Whereas the representational/computational framework seemed geared to describing agent-side processing, the dynamical systems constructs apply as easily to environmental features (e.g., the rhythms of a dripping tap) as to internal information processing events. It is this easy ability to describe larger integrated systems which leads theorists such as Beer and Gallagher (1992) and Wheeler (1994) to prefer Dynamical Systems Theory over classical homuncular approaches for the explanation of emergent, often environment-involving types of behaviors. Behaviors so far studied tend to be relatively basic ones such as legged locomotion and visually guided motion. But the intuition of many such theorists is that the bulk of daily, biological intelligence is rooted in canny couplings between organisms and specific task-environments, and thus that this style of explanation may extend well beyond accounts of relatively "low-level" phenomena. Indeed, Port and van Gelder (1995) contains examples of Dynamical Systems Theorizing applied to such high level tasks as planning and decision making, language production, and event recognition.

Notice, however, that the system parameters tracked in dynamical systems style explanations can be arbitrarily far removed from facts about the real internal structure and processing of the agent. The parameters which figure in a good dynamical systems explanation can thus sometimes *target* values of uncontrolled, or emergent, variables. Thus Tim van Gelder notes that a dynamical systems style story which tracks the behavior of a car engine over time might need to fix on a parameter such as temperature, which does not correspond to any internal component (for homuncular explanation) or to any directly controlled variable. This can occur because:

> In its pure form, dynamical explanation makes no reference to the actual structure of the mechanism whose behavior it is explaining. It tells us how the values of the parameters of the system evolve over time, not what it is about the way the system itself is constituted that causes those parameters to evolve in the specified fashion. It is concerned to explore the topographical structure of the dynamics of the system, but this is a wholly different structure than that of the system itself. (van Gelder, 1991, p. 500)

Intermediate options are clearly also available. Thus Saltzman (1995) offers a Dynamical Systems explanation of how we achieve multiple muscle coordination in speech production. He notes that the coordinative dynamics need to be specified in abstract informational terms which do not directly track either biomechanical or neuroanatomical structure. Instead, "the abstract dynamics are defined in coordinates that represent the configurations of different constriction types, e.g., the bilabial constrictions used in producing /b/, /p/, or /m/, the alveolar constrictions used in producing /d/, /t/, or /n/ etc." (Saltzman, 1995, pp. 159–160.) These constriction types are defined in physical terms involving items such as lip aperture and lip protrusion. But the dynamical systems story is defined over the more abstract types mentioned above. This is an intermediate case insofar as it is clear how the more abstract parameters cited in the dynamical systems analysis are related to physical structures and components of the system. Such intermediate-level analyses are of great importance. Cognitive Science, I shall next argue, cannot afford to do without *any* of the various explanatory styles just reviewed. And it is therefore crucial that we ensure that the various explanations somehow interlock and inform one another. In the second part of this chapter I develop an argument for this explanatory liberalism, and show how the requirement of explanatory interlock itself imposes powerful additional constraints on our theorizing.

4. Two Explanatory Projects

Homuncular explanation and interactive explanation are both well suited to explaining adaptive behavior by unraveling the contributions of specific agent-side components. Interactive explanation differs merely in its explicit recognition of the profound differences which attention to environmental opportunities and the demands of real-time action can make to our hypotheses concerning the information-processing organization required. The Dynamical Systems approach to explaining emergent phenomena, by contrast, imports a whole new perspective, one which focuses on the evolution of system parameters and is especially well suited to modeling the complex interplay between multiple agent-side parameters and environmental ones. Thus described, it seems almost *obvious* that both types of explanation (the homuncular-style analysis and the global dynamics-style analysis) are needed and should be required to interlock gracefully. Yet several recent writings suggest an alternative, more imperialist point of view. Dynamical Systems Theory, they suggest, is to be *preferred over* talk of homuncular decompositions and internal components which traffic in representations. Such a radical view, I shall next argue, can be sustained only by adopting an unduly impoverished vision of the *goals* of Cognitive Science.

We can begin by noticing that fans of Dynamical Systems Theory often treat homuncular decomposition and representation-invoking explanations as somehow going together. Thus Beer claims that adaptive success often obviates any need for a functional decomposition into representations and modules which manipulate them (see, e.g., Beer, 1995, pp. 143–144). And Wheeler (1994, p. 37) even defines the classical vision of homuncular decomposability (which he rejects as a framework for understanding biological cognition) as "the view that we can compartmentalize a system into a hierarchy of specialized sub systems that (i) solve particular sub-tasks by manipulating and/or transforming representations . . . and (ii) communicate the computed outputs to each other by passing representations."

This intimate linking of representation-invoking stories to homuncular decomposition strikes me as problematic. It seems clear that a system *could* be interestingly modular yet not require a representation-invoking explanation (think of a car engine!), and vice versa (think of a connectionist pattern-associator). Moreover, it seems pretty clear that the brain, at least, is a system which exhibits a great deal of modularization and local specialization (see, e.g., Kandel, Schwartz, & Jessell, 2000). As such, it seems ill-advised to define a potential explanatory tool for biological cognition by opposition to homuncular decomposability, at least as the term would naturally be understood.

More importantly, failure to attend to the details of the modular aspects of our internal neural organization would deprive the cognitive scientist of the tools to explain a whole class of phenomena: the systematic effects of various kinds of local damage and disruption. The stress on gross system parameters which helps us understand the dynamics which obtain within whole organism/environment systems must often obscure the details of how different inner systems contribute to that coupling, and thus how the failure of such systems would affect overall behavior. To this end, a great deal of work in Cognitive Neuroscience aims to plot precisely the inner organization which explains patterns of breakdown following local damage (see, e.g., Farah, 1990; Damasio & Damasio, 1994). Such explanations typically adopt both a modular/homuncular and a representation invoking perspective. This kind of understanding *complements* any broader understanding of global dynamics. Each explanatory style helps capture a distinct range of phenomena, and helps provide different types of generalization and prediction.

To take one example, Busemeyer and Townsend (1995) present an elegant application of Dynamical Systems-style theorizing to understanding decision making. The framework they develop, called Decision Field Theory, describes how preference states evolve over time. They describe dynamical equations which plot the interplay of various gross factors (such as the long-term and short-term anticipated value of different choices) and which also predict and explain the oscillations between likely choices which occur during the deliberation process. These are explained as effects of varying the attention which the decision-maker is currently giving to different factors. The account captures and explains several phenomena of interest, including the apparent inconsistencies between preference orderings measured by choice and those measured by selling price (Lichtenstein & Slovic, 1971; Busemeyer & Townsend, 1995). A whole class of generalizations, explanations, and predictions thus falls out of the specific equations they use to model the evolution of the chosen parameters over time.

Other kinds of explanation and generalization, however, are not subsumed by this level of description. Thus consider the famous case of Phineas Gage. Gage was a railway construction foreman in the mid-nineteenth century who suffered a terrible injury when a tamping iron was thrust right through his face, skull, and brain. Amazingly, Gage survived and regained all of his logical, spatial, and physical skills. Memory and intelligence were not affected. Yet after the accident, his life and personality changed dramatically, and for the worse. He was no longer trustworthy, or caring, or able to fulfill his duties and commitments. The damage to his brain had caused, it seemed, a very specific yet strange effect – almost as if his "moral centers" had been destroyed. More accurately, it seemed that his ability to "make rational decisions in personal and social matters" had been selectively compromised, leaving the rest of his intelligence and skills intact. More recently, a team of neuroscientists specializing in the use of brain-imaging techniques analyzed Gage's skull and were able, using computer-aided simulations, to identify the probable sites of neural damage (Damasio et al., 1994). These studies (conjoined with evidence from a group of 12 other patients, including the well-documented contemporary case of the patient EVR; Damasio, Tramel, & Damasio, 1990) offer what A. Damasio has described as "compelling evidence that the human brain has a specialized region for making personal and social decisions" (Blakeslee, 1994; Damasio et al., 1994.)

Such a hypothesis needs careful elucidation since such specializations may nonetheless involve, in the normal case, the conjunctive activation of multiple neural subsystems. Even so, the selective isolation of such networks of networks provides a grain of analysis not present in the more global depiction offered by Decision Field Theory.

The point, for current purposes then, is just that the account of the dynamics of decision-making offered by Decision Field Theory is quite clearly not designed either to predict or illuminate the kind of unexpectedly *selective* disturbance to the decision-making process which the neuroanatomically motivated studies address. This is not a criticism of Decision Field Theory. On the contrary, the latter likewise provides a type of understanding, prediction, and explanation which the former does not. This is because Decision Field Theory is free to treat emergent properties of the whole, intact, well-functioning system as individual abstract parameters, and hence provides a vocabulary with a level of analysis well suited to capturing patterns in the temporally evolving behavior of well-functioning agents. Moreover, it is these more abstract parameters which will often serve us best if we seek to understand the couplings which obtain between whole systems and their environments. Uncontrolled (emergent) parameters, as we saw earlier, often play a large role in illuminating accounts of such large-scale couplings. The moral, then, is that the two styles of explanation are naturally complementary. There is no need for the kind of competition which some of the fans of dynamical systems analysis seem to encourage. Instead, we should clearly distinguish two explanatory projects, each having its own associated class of generalizations. One project aims to understand the way agents and environments are coupled, and may invoke abstract, emergent parameters in so doing. The other seeks to understand the role of various inner subsystems in producing behavior, and hence helps explain whole classes of phenomena (e.g., the effects of local damage) which the other does not address.

5. Microdynamics, Convergence Zones, and the Importance of Explanatory Interlock

One natural way to think of the two projects just outlined is to depict the componential analysis as providing (in part) a story about the implementation of the more global and abstract dynamical systems story. Van Gelder (1991) is skeptical about the value of such implementational stories, at least as regards the understanding of complex neural networks. He notes (p. 502) that componential (or as he says, "systematic") explanation is of little help in cases where "the 'parts' of the structure are so many and so similar, and key parameters . . . do not refer to parts of the system at all."

While this may be true for understanding the behavior of individual and relatively unstructured connectionist networks, it seems manifestly untrue for understanding the brains of most biological organisms. A more realistic picture, I believe, would countenance three equally important and interlocking types of explanation and description:

1. Some account of the gross couplings between the well-functioning organism and the environment. Possibly achieved via Dynamical Systems styles of description, including reference to parameters which are emergent at the level of the individual organism.
2. An account which identifies the neural components whose cooperative behavior is tracked by the parameters mentioned in (1), and which explains (perhaps again using dynamical systems tools) how these components are *themselves* coupled (a kind of micro-level dynamic systems story).
3. At least a coarse-grained analysis of the *types of roles* played by the components identified in (2). It is this last analysis which is most likely to assign representational roles to the activities of individual neural subsystems.

Satisfying explanations of embodied, embedded adaptive success must, I claim, touch all three bases (for a worked-out example, see Clark (1997a, ch. 7, sect. 7.4). Moreover, each type of explanation imposes constraints and requirements on the others. There can be no legitimate agent-side parameters in (1) which lack micro-dynamic implementation detail in (2). And such detail cannot be fully understood without the gross systems level commentary on the *roles* of the various components provided by (3).

6. Conclusions: The Brain Bites Back

A full account even of embodied, embedded, and emergence-laden cognition must do justice to several kinds of data. One important body of data concerns changes in gross system parameters over time. Another concerns the specific effects of local systems damage. Complex evolved biological intelligences seem to rely on a mixture of componential specialization overlaid by increasingly complex systemic properties of feedback and feed-forward modulation. These complex interrelations are probably evolution's way of pressing added utility from existing dedicated applications. The Convergence Zone hypothesis provides a neat example case, in so far as it posits a set of well-defined basic processing areas which are then orchestrated (via the Convergence Zones) into temporary ensembles which invite rather different kinds of description, for example, as supporting retrieval of a concept or of a memory of a unique event.

When evolved systems exhibit this two-sided profile (of componential specialization alongside complex internally interactive properties), the theorist should likewise be willing to exploit multiple kinds of explanatory tools, ranging from analyses which crisscross the organism and the environment, to ones which quantify over multiple inner components and complex connectivities, to ones which isolate components and offer a functional/representational commentary on their basic roles.

Emergent properties will thus figure in this explanatory activity at two levels. First, there will be *internally* emergent features: uncontrolled parameters constituted by the interaction of multiple inner sources of variation. Second, there will be *behaviorally* emergent features: uncontrolled parameters constituted by interactions between whole functioning organisms and the environment. Both classes of emergent property need to be understood, and Dynamical Systems Theory provides a set of tools which can help in each arena. But importantly, these multiple explanatory endeavors are not autonomous. Uncontrolled parameters must be cashed out in real (neural and environmental) sources of variance. And basic componential specialization must be identified and incorporated wherever it occurs. Failure to do so will result in explanatory failure further down the line, for example, when confronted by data concerning the selective impairments caused by local brain damage. In the final analysis, the moral is simple and familiar. The recent emphasis on behavioral and internal emergence is timely and important: but it should not be seen as yet another excuse to avoid confrontation with the biological brain.

Acknowledgment

Parts of this chapter were published earlier in M. A. Boden (ed.), *The Philosophy of Artificial Life*. We thank Oxford University Press for permission to republish these sections.

References

Abraham, R., & Shaw, C. (1992). *Dynamics: The Geometry of Behavior*. Redwood, CA: Addison-Wesley.

Ballard, D. (1991). Animate vision. *Artificial Intelligence*, *48*, 57–86.

Bechtel, W., & Richardson, R. (1992). *Discovering Complexity: Decomposition and Localization as Scientific Research Strategies*. Princeton: Princeton University Press.

Beer, R. D. (1995). Computational and dynamical languages for autonomous agents. In R. Port & T. van Gelder (Eds.), *Mind as Motion: Explorations in the Dynamics of Cognition* (pp. 121–147). Cambridge, MA: MIT Press.

Beer, R., & Gallagher, J. C. (1992). Evolving dynamical neural networks for adaptive behavior. *Adaptive Behavior*, *1*, 91–122.

Blakeslee, S. (1994). Old accident points to brain's moral center. *New York Times*, May 24.

Busemeyer, J., & Townsend, J. (1995). Decision field theory. In R. Port & T. van Gelder (Eds.), *Mind as Motion: Explorations in the Dynamics of Cognition* (pp. 101–120). Cambridge, MA: MIT Press.

Churchland, P. S., Ramachandran, V. S., & Sejnowski, T. J. (1994). A critique of pure vision. In C. Koch & J. Davis (Eds.), *Large-Scale Neuronal Theories of the Brain* (pp. 23–60). Cambridge, MA.: MIT Press.

Clark, A. (1997a). *Being There: Putting Brain, Body and World Together Again*. Cambridge, MA: MIT Press.

Clark, A. (1997b). The dynamical challenge. *Cognitive Science*, *21*, 461–481.

Damasio, A., & Damasio, H. (1994). Cortical systems for retrieval of concrete knowledge: The convergence zone framework. In C. Koch & J. Davis (Eds.), *Large-Scale Neuronal Theories of the Brain* (pp. 61–74). Cambridge, MA.: MIT Press.

Damasio, A., Tramel, D., & Damasio, H. (1990). Individuals with sociopathic behavior caused by frontal damage fail to respond autonomically to social stimuli. *Behavioral Brain Research*, *4*, 81–94.

Damasio, H., Grabowski, T., Frank, R., Galaburda, A., & Damasio, A. (1994). The return of Phineas Gage: Clues about the brain from the skull of a famous patient. *Science*, *264*, 1102–1105.

Dennett, D. (1978). *Brainstorms.* Brighton: Harvester Press.

Farah, M. (1990). *Visual Agnosia.* Cambridge, MA: MIT Press.

Fogel, A. (1993). *Development through Relationships: Origins of Communication, Self and Culture.* Chicago: University of Chicago Press.

Hallam, J. C. T., & Malcolm, C. A. (1994). Behaviour, perception, action and intelligence: The view from situated robotics. *Proceedings of the Royal Society of London, Series A, 349*, 29–42.

Harvey, I., Husbands, P., & Cliff, D. (1994). Seeing the light: Artificial evolution, real vision. In D. Cliff, P. Husbands, J. A. Meyer, & S. Wilson (Eds.), *From Animals to Animats* 3 (pp. 392–401). Cambridge, MA.: MIT Press.

Haugeland, J. (1981). Semantic engines: An introduction to *Mind Design.* In J. Haugeland (Ed.), *Mind Design: Philosophy, Psychology, Artificial Intelligence* (pp. 1–34). Cambridge, MA: MIT Press.

Hofstadter, D. (1985). *Metamagical Themas: Questing for the Essence of Mind and Pattern.* Harmondsworth: Penguin.

Kandel, E. R., Schwartz, J. L., & Jessell, T. (Eds.). (2000). *Principles of Neural Science.* New York: McGraw-Hill.

Kirsh, D. (1995). The intelligent use of space. *Artificial Intelligence, 73*, 31–68.

Le Cun, Y., Boser, B., Denker, J. S., Henderson, D., Howard, R., Hubbard, W., & Jackal, L. (1989). Back propagation applied to handwritten zip code recognition. *Neural Computation, 1/4*, 541–551.

Lichtenstein, S., & Slovic, P. (1971). Reversals of preference between bids and choices in gambling decisions. *Journal of Experimental Psychology, 101*, 16–20.

Maes, P. (1994). Modeling adaptive autonomous agents. *Artificial Life, 1*, 135–162.

Marr, D. (1982). *Vision.* San Francisco: Freeman.

Mascolo, M. F. & Harkins, D. (1998). Toward a component systems approach to emotional development. In M. F. Mascolo & S. Griffin (Eds.), *What Develops in Emotional Development?* (pp. 189–218). New York: Plenum.

Newell, A., & Simon, N. (1976). Computer science as empirical enquiry: Symbols and search. *Communications of the Association for Computing Machines.* Reprinted in M. A. Boden (Ed.), *The Philosophy of Artificial Intelligence* (pp. 104–132). Oxford: Oxford University Press, 1990.

Norton, A. (1995). Dynamics: An introduction. In R. Port & T. van Gelder (Eds.), *Mind as Motion: Explorations in the Dynamics of Cognition* (pp. 45–68). Cambridge, MA: MIT Press.

Port, R., & van Gelder, T. (1995). *Mind as Motion: Explorations in the Dynamics of Cognition.* Cambridge, MA: MIT Press.

Resnick, M. (1994a). *Turtles, Termites and Traffic Jams: Explorations in Massively Parallel Microworlds.* Cambridge, MA: MIT Press.

Resnick, M. (1994b). Learning about life. *Artificial Life*, *1*, 229–242.

Ross, D. (2004). Meta-linguistic signalling for coordination amongst social agents. *Language Sciences*, *26*, 621–642.

Saltzman, E. (1995). Dynamics and co-ordinate systems in skilled sensorimotor activity. In R. Port & T. van Gelder (Eds.), *Mind as Motion: Explorations in the Dynamics of Cognition* (pp. 149–173). Cambridge, MA: MIT Press.

Steels, L. (1994). The artificial life roots of artificial intelligence. *Artificial Life*, *1*, 75–110.

van Gelder, T. (1991). Connectionism and dynamical explanation. *Proceedings of the 13th Annual Conference of the Cognitive Science Society* (pp. 499–503). Hillsdale, NJ: Erlbaum.

Wheeler, M. (1994). From activation to activity. *Artificial Intelligence and the Simulation of Behaviour (AISB) Quarterly*, *87*, 36–42.

Wheeler, M. & Clark, A. (1999). Genic representation: Reconciling content and causal complexity. *British Journal for the Philosophy of Science*, *50*, 103–135.

Wimsatt, W. C. (1986). Forms of aggregativity. In A. Donagan, N. Perovich, & M. Wedin (Eds.), *Human Nature and Natural Knowledge* (pp. 259–293). Dordrecht: Reidel.

Wimsatt, W. C. (2000). Emergence as non-aggregativity and the biases of reductionism(s), *Foundations of Science*, *5*, 269–297.

11

IS PSYCHOLOGICAL EXPLANATION BECOMING EXTINCT?

Cory D. Wright

1. Intertheoretic Reduction versus Reductive Explanation

The doctrine of psychoneural reductionism has long been construed as ultimately rooted in traditional issues in metaphysics and philosophy of science; this is perhaps no more manifest than in attempts to reformulate the mind/body problem as a special application of intertheoretic relations between formally structured pairwise theories replete with lawlike generalizations. Construed thus, proponents of the doctrine sharply distinguished *theory reduction* from *reductive explanation* (Achinstein, 1984; Brandon, 1984; Churchland, 1986; Bickle, 1998, 2003; Barendregt & van Rappard, 2004; Wright & Bechtel, 2006), leaving them to restrict their efforts to the rational reconstruction of these idealized or "textbook" relations using formal set-theoretic analyses. Reductive explanation, on the other hand, could be shuffled off onto historians and sociologists of science, or those whose interests led them to dwell on social, pragmatic, or purely methodological issues involved in the investigation of some target phenomenon Φ. After all, "reductive explanation" – like "explanation" more generally – designates a *communicative practice* between groups of cognizers (be it a butcher, baker, or candlestick maker (or neuropharmacologists)).[1] And mulling over the personal habits and discursive quirks of various researchers hardly seemed a task worthy of the metaphysician or philosopher of science.

The distinction between theory reduction and reductive explanation – paralleling the old "logic of justification" versus "discovery" distinction – afforded psychoneural reductionists a means of compartmentalizing irreducibility claims from the "antireductionist consensus": the *informal explanatory patterns and practices* for some Φ might be carried out independently of lower-level research as long as that independence ultimately didn't carry over to *formal intertheoretic* contexts *as irreducibility*. Such autonomy/independence wasn't threatening to intertheoretic reductionists, since autonomous methods and explanatory practices are crucial to scientific progress, and since the methods and day-to-day explanations inherent in scientific research aren't exhaustively reductive anyway. Bickle, following Churchland (1986, p. 381), effectively makes this type of reply to Pylyshyn:

> One can predict an intertheoretic reduction without tying one's methodological practices to reductive explanations. Reductive *explanations* seek to explain *phenomena* described in a higher-level vocabulary as explananda by exclusive appeal to lower-level kinds and laws as explanans. . . . An inter*theoretic* reductionist can agree wholeheartedly with this methodological point. He need have no commitment to the exclusive use of reductive explanation. . . . Reduction is a prediction about how candidate *theories* will comport with one another *after the fact* of maturity and development. (1998, pp. 153–154)

With this distinction, psychoneural reductionists could keep irreducibility claims at bay in cases where mature or developed theories are reductively related, whilst making the harmless concession that higher-level psychological research has a "useful and seemingly ineliminable heuristic role to play" (Churchland, 1986; Bickle, 1998, p. 153; 2003, p. 130; Craver, 2001).[2] Both in-principle irreducibility of pairwise theories *and* the exclusive use and import of reductive explanatory practice could therefore be rejected *without* inviting charges of inconsistency.

I think that this concession is both sensible and warranted, for broadly Feyerabendian reasons: intertheoretic contexts aside, explanatory practice is hardly *exhausted* by decomposition or other reductive heuristics – crucial and necessary though they are (Wright, 2002). For Bickle, Kim, and other reductionists, this concession apparently conceded too much.[3] Explanatory autonomy/independence of reductive explanation are now also deemed impermissible: for any given Φ, when sufficient amounts of lower-level achievements allow for a successful reductive explanation to obtain, further explanatory contributions from higher-level research are thereby *precluded* because they aren't able to continue offering genuinely penetrating causal insights about the phenomena within their scope. The effect of preclusion on higher-level explanations issuing from such research would seem to be *extinction*. Call this supposition "explanatory extinction" (*EE*).[4] In this chapter, I argue that appeals to actual scientific practice show that higher-level psychological explanations aren't precluded by successful reductive explanation, much less rendered extinct.

2. Psychological Explanation as Fossil Record?

To be clear, *EE* is a supposition about the consequences of successful reductive explanations on higher-level explanatory patterns and practices. *EE* says that the various discursive and explanations typifying the psychological sciences (e.g., adaptive, cognitive, computational, functional) are *de facto* expunged from ongoing research once they discharge the task of enabling increasingly accurate neurobiological explanations; the advent of accomplished finer-grain stories will sanction the act of jettisoning the then-superfluous higher-level stories, like the scaffolding on a finished building. Of course, reductionists can certainly allow, expect, and welcome results from cognitive and clinical neuroscience, experimental psychology, and phenomenology that enable and guide lower-level sciences by, e.g., helping to locate and determine the bounds of the *explanandum*, supplying their own data about higher-level concepts and categories, and providing the necessary initial approximations. In that sense, reductionists have no truck with principles of mutual co-evolutionary feedback and level-bridging; but once a successful reductive explanation obtains, the psychological explanations involved are marked for extinction (Schweizer, 2001; Bickle, 2003, 2006).

There's perhaps no more able a reductionist than Bickle to cite with respect to *EE*, who is worth quoting at length on the matter:

> [W]hen we fix our gaze on aspects of scientific practice in [actual case studies], we see that psychological explanations lose their initial status

> as causally-mechanically explanatory vis-à-vis an accomplished cellular/molecular explanation . . . Within scientific practice, psychological explanations become otiose when the cellular/molecular explanation . . . is achieved. There is no need to evoke psychological causal explanations, and in fact scientists stop evoking and developing them once real neurobiological explanations are on offer. Philosophers who deny this point are usually guided by outdated accounts of real neuroscientific practice. Contra Kim, lower-level explanations need not "exclude" higher-level accounts in any deep epistemological or metaphysical sense. But the former do render the latter pointless, along with any further search for empirically improved successors at the same level – except for some residual, purely heuristic tasks. [A]ccomplished lower-level mechanistic explanations absolve us of the need in science to talk causally, or investigate further, at higher levels, at least in any robust "autonomous" sense. To articulate and defend these claims, I propose that we let scientific practice be our guide. In light of this existing cellular/molecular explanation and these experimental results, it seems silly to count psychology's "explanation" of consolidation as "causally explanatory", "mechanistic", or a viable part of any current scientific investigation worth pursuing. (2003, pp. 110–112)

Plainly, Bickle here asserts *EE*, and avers that failing to do so results from spending too much time gawking at fossil records. The idea, then, is straightforward. When successful reductive explanations obtain, the psychological explanations involved are rendered "pointless" or "silly" curios whose significance becomes entirely historical. The very achievement of successful reductive explanations demands that psychological explanations cede way; and once precluded, they inevitably lose – never to regain – their status as genuinely explanatory pieces of the causal story.

3. Bickle's Wager

If *EE* were true, it would certainly encourage the doctrine of psychoneural reductionism. But is it? And is there an argument for it's being true? What might such an argument look like? In Bickle's assertion of *EE*, the basis for thinking that it's true is not so much an argument as it is a *wager* that successful reductive explanations inevitably *infantilize* higher-level explanatory contributions. The reductionist bets that, by attending to current scientific practice, one will *just see that* such contributions are precluded whereupon a successful reductive explanation obtains.

For the sake of a name, call this wager "Bickle's wager." (To be clear, the target of criticism is neither Bickle himself nor his versions of psychoneural reductionism, but reductionists' deployment of what I am calling "Bickle's wager" to establish the truth of *EE*.) Ironically, Bickle's wager embodies the old Wittgensteinian injunction, "Look, don't think!" Its characteristic devil-in-the-details approach – which fixates on the so-called "put up or shut up" challenges that have recently populated the reductionism literature – proposes that *mere attention to* experimental results in neuroscience will likely reveal accomplished reductive explanations. Philosophers need only look at what's being done in labs across the world, because what's being done just is the reductive explanation of psychological phenomena in terms of explanations of phenomena at much lower levels.

True to form, Bickle antes up with an impressive – though by no means uncontroversial – case study (Schouten & Looren de Jong, 1999; Bickle, 2003, 2006; Craver, 2003). Higher-level psychological explanations of memory, which appeal to the conversion of short-term memory traces to stable long-term memories through stimulus repetition and in the absence of retrograde interference, have been extinguished because of successful reductive explanations of the memory consolidation switch in associative learning theories to the complex mechanisms producing early- and late-phase long-term potentiation (E- and L-LTP). Neuroscientists have shown that this functionally construed posit is subserved by a series of mechanistic activities in CA1 and CA3 regions of the hippocampus (HC). Hence, what was previously defined as a "consolidation switch" at the level of psychological function is now understood as increased synaptic potentiation at cellular/molecular levels; the production of robust excitatory post-synaptic potentials (EPSPs) in mnemonically dedicated circuits results from increased gene expression – specifically, the transcription of genes expressing CREB proteins. Increased gene expression, in turn, results from the activation of protein kinase A and the release of its subunits (which turns off the inhibition of phosphorylated calcium-calmodulin kinase II (CaMKII)), and the conversion of second messenger systems such as cyclic adenosine monophospate (cAMP). For reductionists making Bickle's wager, the larger point is that (putatively) successful reductive explanations like this one can be compiled and parlayed into a strong induction for the doctrine of psychoneural reductionism.

Under some descriptions, this devil-in-the-details fixation is surely worthy of applause. Put innocently enough, there's no good reason not to concur: e.g., "Some might find it difficult in the abstract to see how mind reduces to molecules; but one will never get over that intellectual hurdle or show conclusively why such reductions can't obtain by remaining ignorant of the best existing scientific attempts to do exactly this" (Bickle, 2003, p. 95). But its more extreme forms have nontrivial consequences that should distress philosophers of psychology or cognitive science. For it relegates their role to that of being little more than "science journalists" – i.e., merely reacting to late-breaking neuroscientific developments, and then rehearsing whatever inferences about the mind neuroscientists happen to present them with (albeit in a more popular vernacular). In such extreme forms, the philosophy in "neurophilosophy" is sloughed off along with like-minded disciplines. Consider, e.g., recommendations to bypass research appearing in journals such as *Brain and Behavioral Sciences*, *Journal of Cognitive Neuroscience*, and *Psychological Review* (Bickle, 2006); such eliminativist catcalls portend a revision to the relevant bit of Hume's *Inquiry Concerning Human Understanding*: "Does it contain any (cellular/molecular neuroscientific) experimental reasoning concerning matter of fact and existence? No. Commit it then to the flames, for it can contain nothing but sophistry and illusion."

Bickle's wager requires reviewing experimental results from the best of what science has to offer. So, in Section 4, I'll do just that. I'll review an exemplary case study from behavioral neuroscience – namely, the reductive explanation of an aspect of reward function in terms of dopaminergic operations of the mesocorticolimbic system. If Bickle's wager were a good bet, this successful reductive explanation would establish *EE*. But it doesn't, and *modus tollens* gives you the rest: psychological explanation isn't in danger of going extinct.

4. Mediation of Incentive Salience by Mesocorticolimbic Dopamine

The concept of REWARD falls at the heart of a class of higher-level functional and intentional constructs (desire, appetitive/consummatory pleasure, craving, arousal, reinforced learning, etc.). At least two main features structure this class. First, these constructs are said to possess *direction*, insofar as they are involved in the selection, initiation, and performance of behaviors toward the achievement of some end or goal. Second, they are said to possess *hedonic valence*, insofar as the states and processes they pick out are linked to positive/negative affective values. Reward function exemplifies both of these features; it makes available to an animal a basic functional capacity to be attracted to certain stimuli and to move on the basis of such attraction, and its dysfunction involves characteristic abnormalities in attraction. And so, like most all constructs in this class, it concerns the relationship between an animal's neuropsychobiological profile and its behavior, i.e., those states and processes responsible for its goal-directed commerce in specific environments. Obviously, much more needs to be said about REWARD, what it represents, what a reward function is and how it interacts with other capacities like attention or motivation, whether REWARD can be accurately characterized by causal role functionalism, etc. (In some sense, answers to these questions are precisely what psychoneural reductionists forego; I'll return to this point again in Section 6.)

Fifty years of converging evidence – from a range of species, particularly rats – indicates that the intersection of the sets of neural circuits in which reward function is realized involves a common substrate in the mesocorticolimbic system. Generally, this system – when conjoined with the mesostriatal system – tightly "binds" the directive and hedonic capacities of motivation and pleasure with motor abilities for ascertainment behavior. And dopamine (DA) is one of the main components. Around 80 percent of DA neurons are synchronically activated in mechanisms producing reward (Schultz, 1998), and pharmacological blockade with DA antagonists is well known to induce reward-functional impairments. Within the mesocorticolimbic system, the connectivity between the ventral tegmental area (VTA) and nucleus accumbens (NAc) – 85 percent of which is composed of DA neurons – has garnered much attention as the most significant pathway-governing reward function (Heimer et al., 1991; Fiorino et al., 1993; Pfaus et al., 1995; Ikemoto & Panksepp, 1999; Salamone & Correa, 2002; Hjelmstad, 2004; Wise, 2004).

But how, exactly, do the operations of DA neurons that factor in the activities of the mesocorticolimbic system govern reward functional

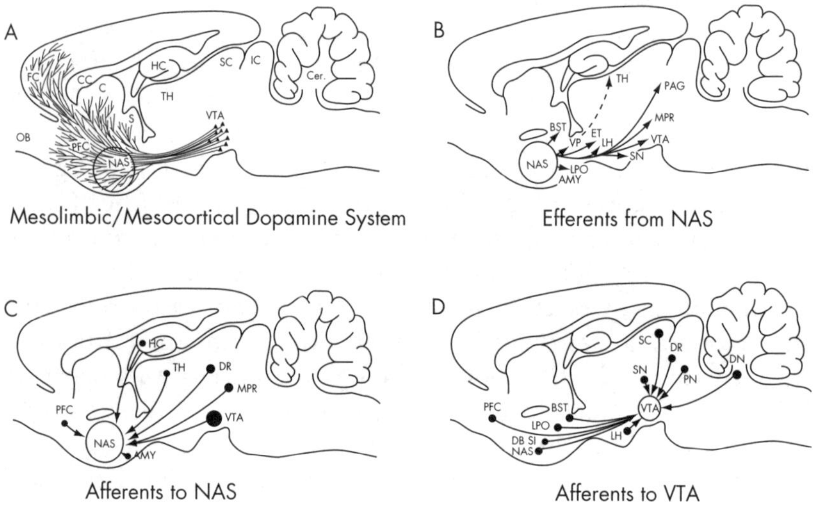

Figure 11.1 Schematic diagrams of mesocorticolimbic pathways in the rat brain
Note: NAS = Nucleus Accumbens, VTA = ventral tegmental area, LH = lateral hypothalamus, and HC = hippocampus.
Source: Reprinted from S. Ikemoto and J. Panksepp, "The role of nucleus accumbens dopamine in motivated behavior," *Brain Research Reviews, 31* (1999): 9, with permission from Elsevier.

states? A now-standard explanation proposes that rewards preferentially potentiate DA transmission in the shell of the NAc (Pfaus et al., 1995; Wise, 1998, 2004; Di Chiara, 1999; Spanagel & Weiss, 1999). Under baseline conditions, DA neurons tonically emit action potentials at low frequencies (≈5–10 Hz), but emit short and sudden *phasic bursts* of high-frequency activity (≈15–20 Hz) that become more prevalent in response to rewarding stimuli – especially those that are *salient* or *novel* (Freeman, Meltzer, & Bunney, 1985; Schultz, 1998; 2002; Casassus, Blanchet, & Mulle, 2005). Functionally, this requires the joint coordination of additional perceptual, attentional, and motivational activities. For example, Horvitz, Stewart, & Jacobs (1997) monitored isolated EPSP waveforms of feline VTA DA neurons in response to visual and auditory stimuli, and discovered that the mere presentation of stimuli dramatically increased the probability of phasic bursting within 200 ms and returned to tonic baseline firing immediately thereafter. By nonlinearly increasing these burst rates, DA neurons can flood a synaptic cleft with "transients" – rapid surges in extracellular DA concentrations that are unachievable in the same time frame with just the same number of normal, low-frequency activation potentials. The failure of expected rewards results in activation patterns that fall back to single spikes (Schultz, 2002). VTA DA neurons, then, provide signals for distinguishing between non-/rewarding stimuli by generating and altering activation patterns induced by phasic bursting, which increase extracellular DA release at the NAc; as they do so indiscriminately across various types of natural and artificial rewards regardless of sensory modality, this operation seems to be the primary neurochemical event that is causally responsible for many aspects of reward function and positive reinforcement (Kiyatkin, 1995).

DA's role in reward function has proven rather more subtle: converging evidence provides reason for thinking that DA doesn't directly code for all aspects thereof, and that many other components and operations are involved (Berridge & Robinson, 1998; Ikemoto & Panksepp, 1999; Spanagel & Weiss, 1999; Berridge, 2003; Wise, 2004; Salamone et al., 2005). Using a DA reuptake blocker and DA_1 receptor antagonist in a probabilistic nosepoke design, Nicola et al. (2005) demonstrated that NAc DA release is both necessary and sufficient for eliciting the causal cascade that eventually produces approach behaviors to rewarding stimuli. Hence, the integrity of the mesocorticolimbic and mesostriatal systems is extremely crucial for being *attracted* by rewarding stimuli; but is DA crucial for coding sheer qualitative pleasure, or in consuming rewarding stimuli? Increases in transient concentrations have been demonstrated by monitoring extracellular changes in DA and DA metabolites (e.g., DOPAC) using *in vivo* microdialysis studies during intracranial self-stimulation of the lateral hypothalamus (LH) (Fiorino et al., 1993; Di Ciano, Blaha, & Phillips, 1998). Self-stimulation is often interpreted as being rewarding because it deceptively activates mesocorticolimbic DA systems that normally code for attractive, novel, or salient stimuli. Yet, Neill, Fenton, & Justice (2002) measured DA and DOPAC concentrations in different LH self-stimulation designs, and found that increased DA firing in the NAc was primarily influenced by the schedules of reinforcement, leading them to conjecture that DA release modulates the response effort and perceived cost of behavior rather than the sheer qualitative pleasure or the actual hedonic properties of stimulation themselves (see Ikemoto & Panksepp, 1996).

Another, more prominent example of subtle refinements to the reductive explanation comes from Berridge and colleagues' studies of unconditioned affective behaviors to food rewards. Based on taxonomies of homologous types of animals' facial expressions and affective behaviors to a variety of tastes and food rewards, they developed an experimental procedure that ostensibly yields observable indications of hedonic valence following manipulations of cellular/molecular DA operations (Berridge, 2000, 2003; Berridge & Robinson, 1998; Robinson & Berridge, 2000, 2004). In one experiment, they induced massive bilateral lesions (>98 percent) of ascending DA neurons from the VTA and DA terminals in the LH and NAc of rats with direct injections of 6-OHDA neurotoxin. If DA operations directly mediated hedonic valence, such total depletions of DA in these key areas should have nullified the ability of rats to experience pleasurable rewards; but neither unconditioned affective behaviors to the positive hedonic valence of sucrose solution nor to the negative hedonic valence of quinine HCl were impaired. (The researchers deduced that this effect wasn't the result of disrupted organization of forebrain and mesocortical systems by the DA lesions, which would be needed for intact associative processes for evaluating the rewarding properties of sweet rewards.) Further, since benzodiazepine agonists enhance positive hedonic valence, then if DA-depleted rats don't suffer from deficits in valence, they should have heightened affective behaviors to sucrose; to confirm this conclusion, diazepam was administered to lesioned rats, which subsequently *did* show the heightened affective behaviors. Accordingly, Berridge and colleagues concluded that mesocorticolimbic DA operations aren't necessary for normal hedonic evaluations (Berridge & Robinson, 1998, p. 328).

Of interest here is that Berridge and colleagues were led to invoke a higher-level distinction to dissociate between aspects of reward function – what they term "core liking" versus "core wanting" – in order explain that two neural subsystems are involved in mediating two different psychological processes. "Core liking" – which was shown to be mediated by a subsystem of which DA isn't a primary component – refers to the attribution of positive hedonic valence to representations of rewarding stimuli (basically, the process of taking some reward to be pleasurable), and is dissociated from "core wanting," which refers to the subpersonal psychological process of forming and transforming representations of ordinary stimuli such that they become especially salient, "attention-grabbing," or otherwise attractive. This distinction – which can be summarized simply as: "one need not want what they like or like what they want" – reformulates a pair of old folk-psychological notions to help make sense of the functional roles played by mesocorticolimbic DA operations.

The latter transformational process, mediated by VTA DA phasic bursting, is necessary for constructing and processing neural representations that attribute *incentive salience* to rewarding stimuli (Berridge & Robinson, 1998; Ikemoto & Panksepp, 1999; Wise, 2002, 2004; McClure, Daw, & Montague, 2003; Robinson & Berridge, 2004; Salamone et al., 2005). Incentively-salient neural representations serve as intermediary links between the activation patterns induced by phasic bursting of VTA DA neurons at NAc terminal fields, and the higher-level cognitive representations interfacing with various reward functional states more generally. These include other attentional and perceptual, affective, learning and memory, motivational, and motor processes that jointly constitute an animal's abilities to ultimately recognize positive reinforcers *as* having a certain hedonic valence, and to elicit approach and even long-term, goal-directed behaviors. Neural representations of incentive salience – and the DA operations that mediate them – thus have a crucial place within the hierarchy of other states and processes at various levels, which constitutes the mesocorticolimbic system and the final common pathway for a repertoire of flexible approach behaviors.[5]

5. A Rewarding Case of Reductive Explanation?

In formal contexts of theory reduction, psychoneural reductionists have been unable to determine how to precisely locate certain cases on the intertheoretic reduction spectrum (Wright, 2000; Fonseca, 2004). For instance, Churchland concludes that determining when a perfectly smooth reduction obtains is a matter of "the whim of central investigators, the degree to which confusion will result . . . , the opportunities for publicizing [results], cadging grants, and attracting disciples" (1986, pp. 283–284); likewise, Bickle concedes that it's "ridiculously optimistic" to provide formal criteria for distinguishing between perfectly bumpy reductions from eliminations (Bickle, 1998, p. 101). In informal contexts of reductive explanation – where success and sufficient completeness of investigation and understanding is much more nebulous – providing positive criteria is all the more difficult; unfortunately, asserting *EE* incurs the onus of doing just that.

Since it's unlikely that reductionists will have such criteria in informal contexts of reductive explanation, let's simply grant them, for the sake of argument, that the reward function case is a case of successful reductive explanation. This is an eminently plausible assumption. For what I've presented barely scratches the surface of a massive corpus of research on the realization of aspects of reward function. And there's no doubt that epistemic progress has partially resulted from lower-level explanatory pressures, like the resolution of important anomalies. For example, one problem for past explanations of reward function in terms of mesocorticolimbic mechanisms is that well over two-thirds of relevant DA neurons seem to be excited by aversive stimuli (Ikemoto & Panksepp, 1999). Recent electrophysiology and immunofluorescence work by Ungless, Magill, & Bolam (2004) demonstrated that the anomaly is only apparent: DA neurons *are* uniformly inhibited by aversive stimuli because the aversively-excitable neurons weren't actually dopaminergic. Many further refinements still need to be made, details specified, anomalies resolved, more penetrating formulation of constructs, etc.; but even this all-too-brief review exhibits the stable outlines of a successful reductive explanation already hewn. At the very least, psychoneural reductionists would be hard-pressed to deny it. For one thing, the reductive explanation of incentive salience in terms of DA bursting/gating easily rivals that of memory consolidation in terms of E- and L-LTP; hence, denial would therefore put reductionists in the uncomfortable position of seeming hasty or disingenuous about the success of the latter case. Moreover, denial would fail to square with many reductionists' staunch dismissal of epistemic impoverishment claims (i.e., bemoanings of "how little we know about how the brain works"; Bickle, 2003, 94–95); hence, if such claims are indeed gross miscalculations about what is known, then surely this case constitutes a successful reductive explanation if any does (i.e., it's hard to feel sympathetic with the reductionist who tries to have it both ways).

6. The Importance of Being Level-Headed

It's important to note that even successful reductive explanations are susceptible to explanatory refinement: there's always the possibility of compiling more information, of better stories to tell, different ways of telling them, and new connections to forge – even "after the fact of maturity and development." Intuitively, it might appear that susceptibility to explanatory refinement entails that the reductive explanation wasn't actually an accomplished case after all. But this intuition implies that reductive explanations need to be complete to count as successful; and surely such a requirement is too strong, since most reductive explanations would thereby be rendered "unsuccessful." In any case, the appearance of inconsistency in psychoneural reductionist doctrine – particularly, in Bickle's wager – easily dissipates *if* explanatory refinements aren't taken to originate from higher-level research. For recall that anyone making Bickle's wager is only committed to a conditional: if we attend to the details of actual scientific practice, we will find that *EE* holds in cases where accomplished reductive explanations obtain. So consistency dictates only that reductionists assert the further claim that refinements are permissible if they take a "more-of-the-same" form (e.g., making more low-level manipulations and specifying the causal effects on higher-level psychological processes and behavior); hence, *once* a successful reductive explanation obtains, only higher-level research loses its explanatory and revisionary potency. It's this further assertion that should be contested. For starters, if dissolving the appearance of inconsistency requires that explanatory refinements don't originate from higher-level psychological research, then *EE* is a patently question-begging supposition, since that's the very issue under dispute.

The "intervene-cellularly/molecularly, track-behaviorally" strategy purportedly best captures experimental and explanatory practice (Bickle, 2003, 2006); and indeed, reward researchers often utilize, e.g., lesion/deficit preparations to understand which components' absence disrupts normal mechanistic activity, as well as preparations involving excitation and inhibition of lower-level component operations. Lesion experiments with 6-OHDA neurotoxin are a prime example. But while reductionists are quite right to focus on reductive explanations born of these sorts of reductive procedures, the "intervene-and-track" strategy is hardly the only experimental procedure used by researchers to further develop already accomplished explanations. Many researchers also use what are called "contextual" or "additive" strategies (Craver, 2001), such as methods that track the effects of environmental conditions or cues on behavioral paradigms, or that investigate the effects that larger, more encompassing mechanisms have on the component parts in which they operate.[6] For instance, a crucial means of implicating phasic bursting of DA neurons in incentive salience and reward-related learning was to simply chronicle what individual DA neurons do when the animal as a whole is put in a particular environment and presented with different types of distal stimuli (Horvitz et al., 1997). In another single-unit recording study, Taha and Fields (2005) implanted electrode arrays in rats' NAc and then ran the animals through two behavioral paradigms – a sucrose discrimination task and a contrast task to determine whether NAc cells code for hedonic valence. By recording the activities of select neurons and then manipulating the whole animal and its environment, researchers were able to distinguish two operations performed by the neural population: excitations encoding the hedonic valence of sucrose solutions, and disinhibitions of other neurons to initiate and maintain consummatory behaviors. As a matter of public record, these non-reductive, "top-down" strategies provide refinements even *after* successful reductive explanations has obtained; ergo, rather than merely enabling more lower-level explanation, they provide new information based on manipulations of higher-level variables. The new information provided is usually of the sort left out of reductive explanations – namely, information about how functions and mechanistic activities are embedded in more encompassing systems, about functional significance and better role specifications, about the effects of environmental or adaptive contexts in which Φ is situated, better understanding of systemic-level activities, and so forth. After all, VTA DA neurons must be causally hooked up to each other in appropriate ways in order to provide signals that form and transform incentively-salient neural representations. And the circuits they compose must themselves be organized so as to efficiently interact with larger and more complex mechanistic subsystems, which are themselves embedded within larger mesocortical, mesolimbic, and mesostriatal systems. But as increasingly sophisticated explanations then circle back to explain these higher levels of systemic organization – such that the ultimate set of relations between *explanans* and *explanandum* can be understood – the significance of each individual lower-level component cedes explanatory priority (Wright & Bechtel, 2006, p. 61). With this "circling back," the explanatory focus becomes the overall activity of the larger circuits and mechanistic systems comprising the final common pathways affording cognizers the very abilities and capacities to construct nuanced cognitive representations of reward-related learning, behavior, and stimuli.

Of course, contextual and additive strategies – while casting doubt on reductionists' claims that explanatory refinements only originate from lower-level research – aren't yet enough to save psychological explanation from extinction. It must be shown that psychological-level research itself still plays some role, however minimal. Here, examples are more difficult to find – but not impossible. One way that psychological experiments are crucial is in helping evaluate experimental procedures and models against certain kinds of validation criteria – particularly, convergent and discriminant validity, and face validity (Wright, 2002). For instance, Germans and Kring (2000) show that some self-report measures of anhedonia (e.g., Scale for Physical Anhedonia) aren't adequately sensitive to detect differences in affect, but that other measures – such as self-report of approach motivation and motivational salience – may better index hedonic deficits. Such (purely) psychological studies corroborate what is known about the general role of motivation and incentive salience in reward function and provides reason to filter out particular clinical measures that may be obscuring what little human or clinical data is available. The interesting thing about this sort of example is that, *despite* giving an extremely anemic sense of what higher-level psychological research on reward function amounts to, it's nevertheless *strong enough* to show why an understanding of the cellular/molecular mechanisms of reward function, once achieved, doesn't simply leave psychological explanations with *nothing* to do.[7]

Another reason for contesting the question begging assertion that explanatory refinements can't originate from higher-level psychological research is that successful reductive explanations don't simply invoke successively lower-level vocabularies. Communicative practices being what they are, scientific explanations are often constructed using miscegenated vocabularies from different subdisciplines – there simply are no neat divisions among lexical items. This should not be taken lightly. Consider, e.g., the reward function research that's been taken up by the burgeoning field of neuroeconomics (Schultz, 2004). Bridging neuroscientific and economic explanations would hardly be possible were it not for the higher-level psychological constructs involved in explaining reward-related learning, expectation, uncertainty, and decision making. DA operations in reward-related learning, for instance, involve numerous appeals to information-theoretic and cognitive terms ("processes feedback information," "signal," "uncertainty," "evaluating expectations," etc.). Hence, it should be no surprise why the reductive explanation of this aspect of reward functionality nontrivially interprets mesocorticolimbic DA operations using the conceptual apparatus (e.g., PREDICTION, ERROR-RECOGNITION, EXPECTATION) of higher-level information theory (Gallistel, 1994; Cohen, Braver, & Brown, 2002; Schultz, 2004; Casassus et al., 2005). For example, Schultz, Dayan, & Montague (1997; Schultz 1998, 2002) used computational models of temporal difference-learning algorithms to demonstrate that DA firing patterns in the VTA and substantia nigra (SN) obey formulas representing the deviation or error between predictions about the timing, direction, and valence of rewards, and the properties of those rewards which actually obtain. These discriminatory predictions of uncertainty and error vis-à-vis expected hedonic valence are what we might call "protoepistemic processes," and are themselves organized and coordinated such that they factor into larger, more complicated epistemic or cognitive processes. McClure et al. (2003) adapted these same computational models to show how the concept of EXPECTED FUTURE VALUE in Schultz & colleagues' work could be identified with the concept of INCENTIVE SALIENCE ATTRIBUTION in that of Berridge and colleagues'. The result of such bridging was "a more formal computational theory of how [DA] is involved in a larger system for choosing optimal actions under the motivation of prediction errors" (2003, 427; cf. Schweizer, 2001). Suffice it to say that successful reductive explanations invoke a variety of ideas and distinctions originating from numerous levels; Berridge and colleagues' employment of "liking" versus "wanting" to clarify the role of DA in incentive salience attribution nicely exemplifies this. As a blend of folk and scientific-psychological constructs, the distinction allowed researchers to interpret theoretically complex results in terms of a simpler conceptual domain that is already well understood. And as a slice of actual scientific practice, understanding a poorly understood or unwieldy domain in terms of something more familiar or comprehensible is undoubtedly the norm in scientific reasoning. Successful reductive explanations are no different, synthesizing a variety of conceptual structures and relying on interpretations of data derived from models and experimental techniques that integrate phenomena of different grain and particularity.

This last point becomes more prominent still upon noting that neurobiologists themselves constantly make explicit calls for further evidence and constraints from the psychological sciences. In their review, Robbins and Everitt concluded, "Even leaving aside the complications of the subjective aspects of motivation and reward, it is probable that further advances in characterizing the neural mechanisms underlying these processes will depend on a better understanding of the *psychological* basis of goal-directed or instrumental behavior" (1996, p. 228; my emphasis). In particular, they noted an immediate need for functional neuroimaging results in cognitive neuropsychology to help localize large-scale task-relevant information-processing operations. Similarly, Berridge and Robinson aver, "[F]urther advances will require equal sophistication in parsing reward into its specific *psychological* components . . . Neuroscientists will find it useful to distinguish the *psychological* components of reward because understanding the role of brain molecules, neurons, and circuits requires understanding what brains really do – which is to mediate specific behavioral and psychological functions" (2003, p. 507; my emphasis). So, have reductionists shown that "The causal explanations of quantitative behavioral data eschew appeals to concepts at levels higher than cellular/molecular mechanisms and the anatomical circuitries containing them" (Bickle, 2006)? Not remotely. Indeed, the assertion of *EE* now appears awkwardly committed to the further supposition that neurobiologists themselves suffer from a sort of false consciousness or are otherwise deluded. For what's "explicit in scientists' writings" is the claim that better scientific understanding of what's been successfully explained demands *continued* conceptual contributions at higher levels.

The fact that neurobiologists make such calls for developments from psychological sciences *despite* having a wealth of lower-level data to invoke underscores the fact that making sense of the significance of phasic bursting of VTA DA neurons depends on how well illuminated the reward constructs involved are – i.e., the extent to which the process of incentive salience attribution or "core wanting" is understood, what motivation processes are involved, what CRAVING is how appetitive behavior is different from consummatory behavior or approach responses, etc. Since the context here is explanation and not theory reduction, *understanding* is crucial. If the very concept of REWARD, or any of the fragmented constructs from that general class of phenomena, are undercharacterized or conceptually impoverished at the time of successful reductive explanation, then obviously there higher-level developments that still need to be made. This point is made plain by reconsidering how Section 4 began – with an intentionally all-too-brief account of the concept of REWARD, what a reward function is, whether the extensions of "reward" and "positive reinforcement" pick out equivalence classes of isomorphic properties, etc. In some sense, the brevity of such descriptions serves as a litmus test: even after a successful reductive explanation obtains, is further characterization of higher-level constructs needed? I see no good reason for thinking that researchers at higher levels should dispense with the desideratum of self-critically improving on their own psychological explanations. Using the construct of MOTIVATION as his example, here's how Berridge puts the point:

> Motivational concepts are becoming widely recognized as needed to help neuroscience models explain more than mere fragments of behavior. Yet, if our motivational concepts are seriously wrong, our quest for closer approximation to brain-behavior truths will be obstructed as much as if we had no concepts at all. We need motivational concepts, and we need the right ones, to properly understand how real brains generate real behavior. (2004, p. 180)[8]

So, even with a successful reductive explanation of aspects of reward function at hand, there's still a need, desire, craving, drive, urge, hunger, impulse, motivation, incentive, etc., for continued refinements in higher-level characterizations.

Accordingly, neurobiological developments notwithstanding, research on reward function is no different than virtually any other case study in the following respect: purely insular research that deliberately eschews bridges with research at other levels once accomplishments obtain is a rarity. Instead, investigations typically span several different levels – of both organization and analysis – and most experiments utilize various techniques that integrate domains (Wright & Bechtel, 2006). If so, then it stands to reason that the explanations issuing from actual scientific practice cannot but help reflect this multilevel, pluralistic approach. A candid look at the discourse of actual scientific practice reveals why a broadly Feyerabendian analysis is gripping: rather than being restricted to research from increasingly lower levels, dis-/confirming supplementary results potentially come from any direction, from any level, from any branch of science or epistemology. Good explanatory practice just simply exploits that fact.

7. Conclusion

In order to make good on *EE*, the reductionist must demonstrate that actual scientific practice is played out in a certain reductionist key, and that philosophers need only attend to the details in order to hear the tune. I hope to have shown that the details don't quite pan out that way. The obtaining of a successful reductive explanation, born of accomplishments of lower-level research, fails to serve as a condition on the impossibility of achieving better scientific understandings of the target phenomenon Φ at higher levels. Now, reductions – when they occur – are fine so far as they go. Indeed, the construction of models involving the localization of mechanistic activities and the decomposition into their component parts and operations typify some of the ratiocinative process of understanding psychological phenomena (Wright & Bechtel, 2006). But does that entail that detailed attention to actual scientific practice will portend the sort of tell-all reductionist tale that some might wish to see? I see reason for caution. The case study briefly reviewed here suggests that Bickle's wager isn't a safe bet; the doctrine of psychoneural reductionism is best off purging commitments to *EE*, or at least leaving it at the laboratory door.

In successfully carrying out a reductive explanation of aspects of reward function, researchers often conscientiously work toward a unified, integrated story that bridges the levels inherent in the phrase "neuropsychobiological profile" very seriously (Gallistel, 1994; Ikemoto & Panksepp, 1999; Berridge & Robinson, 2003; McClure et al., 2003; Schultz, 2004).[9] Refinements in reductively understanding the target phenomenon obtain partly because many of the fragmented constructs are characterized across various levels of research, not just successively lower ones. And the explanation of those constructs characteristically follows a deeply entrenched pattern of terminological and conceptual miscegenation across various levels. So, while psychological explanations of reward function have often played a crucial heuristic role in enabling increasingly accurate neurobiological explanations, the conjunction of that heuristic role and the increased sophistication of lower-level research does not necessarily entail – at least not in this paragon case – that higher-level psychological explanations of paradigmatic functions are generally rendered "impotent" or are left to "disappear" (cf. Bickle, 2003, p. 111). Often, those making Bickle's wager are quick to dispute the charge that they've misunderstood the scientific details. But I don't think that's the problem at all. Psychoneural reductionists understand those details perfectly well, and are certainly right to hold that more attention to them is a good thing; the problem is that selective attention isn't.

Acknowledgments

This research was supported by a grant from the William J. Fulbright foundation. Conversations about it with William Bechtel, John Bickle, Anthony Landreth, Ioan Muntean, Ken Sufka, Dingmar van Eck, and Iris van Rooij have been extremely rewarding.

Notes

1 At its most general, "psychoneural reductive explanation" designates a communicative practice that involves one group of cognizers employing various (semiological) conventions to prompt another to produce narrowly interpretable conceptualizations about a psychological-level *explanandum* in terms of a neurobiological-level *explanans.* Typically, this involves both a description of the manner in which higher-level research has initially characterized the occurrence, causal powers, or dynamics of the target phenomenon Φ, and the precise spatial, temporal, and organizational properties, relations, and operations of ontological posits as determined by lower-level research.

2 Feest (2003) argues that, while irreducibility may entail the autonomy/independence of psychological explanation, the converse does not follow;

and so reductionists who eschew autonomy because it leads to irreducibility are equivocating between intertheoretic and explanatory autonomy (or, as she puts it, between "explanatory autonomy" and "investigative autonomy"). The argument is plausible, despite the fact that Feest proposes that reductionists are right to claim that psychological explanation fails to be explanatorily autonomous, but wrong to claim that they thereby fail to be intertheoretically autonomous (as the quotation shows, this simply gets the reductionist claim backwards).

3 As always, Bickle has been a forerunner in streamlining the doctrine of psychoneural reductionism – most recently, by turning away from post-classical accounts such as his structuralist New Wave program in order to focus more explicitly on the day-to-day research emerging from cellular/molecular neuroscience. Though he maintains that the formal rendering of the intertheoretic reduction relationship ρ is correct so far as it goes, Bickle's turn toward reductive explanations in cellular/molecular neuroscience is motivated by the recognition that large-scale theories aren't being constructed in actual scientific practice – at least not in the grand sense that was being developed in the New Wave program (personal communication).

4 Note that "extinction" is used as a technical term peculiar to the reductive explanatory patterns inherent in actual scientific practice, and so is *not* synonymous with either "elimination" or "replacement." The latter terms are best reserved for, e.g., intertheoretic contexts where a set of inadequate explanations are unable to be formally reconstructed as part of the set of models and laws comprising some reduced theory T_R.

5 Interestingly, the reductive explanation of incentive salience attribution also provides a basis for helping explain reward function gone awry. Di Chiara (1998, 1999) proposed the chronic stimulation of DA neurons in the NAc shell abnormally strengthens stimulus-drug associations of reward-related learning, which artificially confers excessive incentive salience to stimuli predictively associated with some drugs of abuse (e.g., cocaine). This conjecture is buttressed by hypotheses about the dysphoria and anhedonic experiences common in disorders such as depression and schizophrenia. Kapur (2004) proposes that some aspects of psychotic symptoms, for instance, can also be explained by aberrant spiking of DA neurons independent of cue and context; the context inappropriate release of DA erroneously triggers larger patterns of neurocomputational activation that bring about the confused feelings of altered incentive salience. Over time, the incessant dopaminergic malfunction ends up hijacking normal processes of salience attribution, prediction, and learning. In sum: "Dysregulation of the dopamine system provides the fuel for the creation of the delusion, whereas the patient's personal and cultural

history gives it the precise form" (Kapur, 2004, p. 403). The aberrant salience of psychotic delusions can be attenuated by antipsychotics, whose mechanisms of action include globally damping mesocorticolimbic DA operations (but, unfortunately, thereby downgrades subjects' normal reward function).

6 Since interlevel causation is not necessarily being invoked, *a fortiori* no story about downward causation is required.

7 Some reductionists might be inclined to reply by gesturing at the telling imbalance between higher- and lower-level research (e.g., Bickle, 2003, p. 4). But this would obviously be a red herring: whether neurobiological research on reward function radically "outproduces" its higher-level counterpart is certainly telling, but largely irrelevant to the whether *EE* is true.

8 Motivation constructs are interesting for the case of reward function because they're implicated in a wide range of phenomena – from the intermediate processes involved in incentive salience attribution, to the construction of abstract goals, social ambitions, personal achievement, moral ends, etc. Widening the explanatory scope of reward function research in this way yields yet another reason why contributions from a range of higher-level disciplines are not pointless or silly.

9 Barendregt and van Rappard (2004) welcome the idea of bringing the nature of reductive explanation back into vogue, arguing that the doctrine of psychoneural reductionism has been mired in confusion precisely because of the overemphasis on theory reduction. Instead, they propose that the doctrine is best understood as a general methodological stance which facilitates interlevel cooperation and bridges theories at different levels without ontological concerns. On one hand, what they characterize as "reductive explanation" seems to accurately characterize much of actual scientific practice. On the other hand, what they call "reductive explanation" is actually mechanistic, not reductive, explanation (for explication of the difference between these two types, see Brandon, 1984; Craver, 2001; Wright & Bechtel, 2006). Consequently, if psychoneural reductionist asserting *EE* thereby deny the autonomy/independence of higher-level psychological explanation, this sort of conflation cannot but help fail to make sense of it.

References

Achinstein, P. (1984). The pragmatic character of explanation. *Philosophy of Science*, *2*, S275–S292.

Barendregt, M., & van Rappard, J. F. H. (2004). Reductionism revisited: On the role of reduction in psychology. *Theory and Psychology*, *14*, 453–474.

Berridge, K. C. (2000). Measuring hedonic impact in animals and infants: Microstructure of affective taste reactivity patterns. *Neuroscience and Biobehavioral Reviews*, *24*, 173–198.

Berridge, K. C. (2003). Pleasures of the brain. *Brain and Cognition*, *52*, 106–128.

Berridge, K. C. (2004). Motivation concepts in behavioral neuroscience. *Physiology and Behavior*, *81*, 179–209.

Berridge, K. C., & Robinson, T. E. (1998). What is the role of dopamine in reward: Hedonic impact, reward learning, or incentive salience? *Brain Research Reviews*, *28*, 309–369.

Berridge, K. C., & Robinson, T. E. (2003). Parsing reward. *Trends in Neurosciences*, *26*, 507–513.

Bickle, J. (1998). *Psychoneural Reduction: The New Wave*. Cambridge, MA: MIT Press.

Bickle, J. (2003). *Philosophy and Neuroscience: A Ruthlessly Reductive Account*. Dordrecht: Kluwer.

Bickle, J. (2006). Reducing mind to molecular pathways: Explicating the reductionism implicit in "molecular and cellular cognition." *Synthese*.

Brandon, R. N. (1984). Grene on mechanism and reductionism: More than just a side issue. *Philosophy of Science*, *2*, S345–S353.

Casassus, G., Blanchet, C., & Mulle, C. (2005). Short-term regulation of information processing at the corticoaccumbens synapse. *Journal of Neuroscience*, *25*, 11504–11512.

Churchland, P. S. (1986). *Neurophilosophy: Toward a Unified Science of the Mind–Brain*. Cambridge, MA: MIT Press.

Cohen, J. D., Braver, T. S., & Brown, J. W. (2002). Computational perspectives on dopamine function in prefrontal cortex. *Current Opinion in Neurobiology*, *12*, 223–229.

Craver, C. F. (2001). Role functions, mechanism, and hierarchy. *Philosophy of Science*, *68*, 53–74.

Craver, C. F. (2003). The making of a memory mechanism. *Journal of the History of Biology*, *36*, 153–195.

Di Chiara, G. (1998). A motivational learning hypothesis of the role of dopamine in compulsive drug use. *Journal of Psychopharmacology*, *12*, 54–67.

Di Chiara, G. (1999). Drug addiction as dopamine-dependent associative learning disorder. *European Journal of Pharmacology*, *375*, 13–30.

Di Ciano, P., Blaha, C. D., & Phillips, A. G. (1998). Conditioned changes in dopamine oxidation currents in the nucleus accumbens of rats by stimuli paired with self-administration or yoked-administration of d-amphetamine. *European Journal of Neuroscience*, *10*, 1121–1127.

Feest, U. (2003). Functional analysis and the autonomy of psychology. *Philosophy of Science*, *70*, 937–948.

Fiorino, D. F., Coury, A., Fibiger, H. C., & Phillips, A. G. (1993). Electrical stimulation of reward sites in the ventral tegmental area increases dopamine transmission in the nucleus accumbens of the rat. *Behavioral Brain Research*, *55*, 131–141.

Fonseca, J. (2004). On Bickle's failure to give a formal account of location in the New Wave reductionist spectrum. *Disputatio*, *1*, 50–60.

Freeman, A. S., Meltzer, L. T., & Bunney, B. S. (1985). Firing properties of substantia nigra dopaminergic neurons in freely moving rats. *Life Sciences*, *36*, 1983–1994.

Gallistel, C. R. (1994). Foraging for brain stimulation: Toward a neurobiology of computation. *Cognition*, *50*, 151–170.

Germans, M. K., & Kring, A. M. (2000). Hedonic deficit in anhedonia: Support for the role of approach motivation. *Personality and Individual Differences*, *28*, 659–672.

Heimer, L., Zahm, D. S., Churchill, L., Kalivas, P. W., & Wohltmann, C. (1991). Specificity in the projection patterns of accumbal core and shell in the rat. *Neuroscience*, *41*, 89–125.

Hjelmstad, G. O. (2004). Dopamine excites nucleus accumbens neurons through the differential modulation of glutamate and GABA release. *Journal of Neuroscience*, *24*, 8621–8628.

Horvitz, J. C., Stewart, T., & Jacobs, B. L. (1997). Burst activity of ventral tegmental dopamine neurons is elicited by sensory stimuli in the awake cat. *Brain Research*, *759*, 251–258.

Ikemoto, S., & Panksepp, J. (1996). Dissociations between appetitive and consummatory responses by pharmacological manipulations of reward-relevant brain regions. *Behavioral Neuroscience*, *110*, 331–345.

Ikemoto, S., & Panksepp, J. (1999). The role of nucleus accumbens dopamine in motivated behavior: A unifying interpretation with special reference to reward-seeking. *Brain Research Reviews*, *31*, 6–41.

Kapur, S. (2004). How antipsychotics become anti-"psychotic" – from dopamine to salience to psychosis. *Trends in Pharmacological Sciences*, *25*, 402–406.

Kiyatkin, E. A. (1995). Functional significance of mesolimbic dopamine. *Neuroscience and Biobehavioral Reviews*, *19*, 573–598.

McClure, S. M., Daw, N. D., & Montague, P. R. (2003). A computational substrate for incentive salience. *Trends in Neurosciences*, *26*, 423–428.

Neill, D. B., Fenton, H., & Justice, J. B. (2002). Increase in accumbal dopaminergic transmission correlates with response cost not reward of hypothalamic stimulation. *Behavioral Brain Research*, *137*, 129–138.

Nicola, S. M., Taha, S. A., Kim, S. W., & Fields, H. L. (2005). Nucleus accumbens dopamine release is necessary and sufficient to promote the behavioral response to reward-predictive cues. *Neuroscience*, *135*, 1025–1033.

Pfaus, J. G., Damsma, G., Wenkstern, D., & Fibiger, H. C. (1995). Sexual activity increases dopamine transmission in the nucleus accumbens and striatum of female rats. *Brain Research*, *693*, 21–30.

Robbins, T. W., & Everitt, B. J. (1996). Neurobehavioral mechanisms of reward and motivation. *Current Opinion in Neurobiology*, *6*, 228–236.

Robinson, T. E., & Berridge, K. C. (2000). The psychology and neurobiology of addiction: an incentive-sensitization view. *Addiction*, *95*, S91–S117.

Robinson, T. E., & Berridge, K. C. (2004). Incentive-sensitization and drug "wanting." *Psychopharmacology*, *171*, 352–353.

Salamone, J. D., & Correa, M. (2002). Motivational views of reinforcement: Implications for understanding the behavioral functions of nucleus accumbens dopamine. *Behavioral Brain Research*, *137*, 3–25.

Salamone, J. D., Correa, M., Mingote, S. M., & Weber, S. M. (2005). Beyond the reward hypothesis: Alternative functions of nucleus accumbens dopamine. *Current Opinion in Pharmacology*, *5*, 34–41.

Schouten, M. K. D., & Looren de Jong, H. (1999). Reduction, elimination, and levels: The case of the LTP-learning link. *Philosophical Psychology*, *12*, 237–262.

Schultz, W. (1998). Predictive reward signal of dopamine neurons. *Journal of Neurophysiology*, *80*, 1–27.

Schultz, W. (2002). Getting formal with dopamine and reward. *Neuron*, *36*, 241–263.

Schultz, W. (2004). Neural coding of basic reward terms of animal learning theory, game theory, microeconomics and behavioral ecology. *Current Opinion in Neurobiology*, *14*, 139–147.

Schultz, W., Dayan, P., & Montague, P. R. (1997). A neural substrate of prediction and reward. *Science*, *275*, 1593–1599.

Schweizer, P. (2001). Realization, reduction, and psychological autonomy. *Synthese*, *126*, 383–405.

Spanagel, R., & Weiss, F. (1999). The dopamine hypothesis of reward: Past and current status. *Trends in Neuroscience*, *22*, 521–527.

Taha, S. A., & Fields, H. L. (2005). Encoding of palatability and appetitive behaviors by distinct neuronal populations in the nucleus accumbens. *Journal of Neuroscience*, *25*, 1193–1202.

Ungless, M. A., Magill, P. J., & Bolam, J. P. (2004). Uniform inhibition of dopamine neurons in the ventral tegmental area by aversive stimuli. *Science*, *303*, 2040–2042.

Wise, R. A. (1998). Drug-activation of brain reward pathways. *Drug and Alcohol Dependence*, *51*, 13–22.

Wise, R. A. (2002). Brain reward circuitry: Insights from unsensed incentives. *Neuron*, *36*, 229–240.

Wise, R. A. (2004). Dopamine, learning, and motivation. *Nature Reviews Neuroscience*, *5*, 483–494.

Wright, C. D. (2000). Eliminativist undercurrents in the new wave model of psychoneural reduction. *Journal of Mind and Behavior*, *21*, 413–436.

Wright, C. D. (2002). Animal models of depression in neuropsychopharmacology qua Feyerabendian philosophy of science. In S. P. Shohov (Ed.), *Advances in Psychology Research* (Vol. 13, pp. 129–148). New York: Nova-Science Publishers.

Wright, C. D., & Bechtel, W. (2006). Mechanisms and psychological explanation. In P. Thagard (Ed.), *Handbook of Philosophy of Psychology and Cognitive Science* (pp. 31–79). New York: Elsevier Science.

12

WHO SAYS YOU CAN'T DO A MOLECULAR BIOLOGY OF CONSCIOUSNESS?

John Bickle

1. "Molecular Consciousness Studies"?

The answer to this chapter's title is: Pretty much everybody now working in consciousness studies. This includes philosophers whose attacks stem from global denials about any physical science accounting for some aspects of conscious experience. Colin McGinn is a clear example: "We are cut off by our very cognitive constitution from achieving a conception of that natural property of the brain (or of consciousness) that accounts for the psychophysical link" (1989, p. 350) – molecular biological properties among these. But it also includes less skeptical philosophers and scientists who seek explanations of consciousness at higher levels of biological organization, such as information processing (e.g., Baars, 1988), neural networks (e.g., Paul Churchland, 1995; Patricia Churchland, 2002), and clinical neuropsychology and neurology (e.g., Ramachandran & Blakeslee, 1998; Damasio, 1999).

Even the handful of basic neuroscientists who concern themselves with consciousness say little to nothing in print about potential molecular mechanisms. Francis Crick and Christof Koch's now-classic essay (1990) is a good example. They are not mysterians: "We suggest that time is now ripe for an attack on the neural basis of consciousness" (1990, p. 264). Nor are they cognitivists: "Arguments at the cognitive level are undoubtedly important but we doubt whether they will, by themselves, ever be sufficiently compelling to explain consciousness in a convincing manner" (1990, p. 264). And yet, four paragraphs after these statements, they inform us that "paradoxically, consciousness appears to be so odd and, at first sight, so difficult to understand that only a rather special explanation is likely to work" (1990, p. 265). They provide "a sketch of a theory" involving saliency and feature maps across thalamus and cortex, oscillation frequencies through the anatomical circuits connecting them, and a short-term (iconic, working) memory store. None of this sketch speaks to molecular mechanisms. Admittedly this paper is dated scientifically. But more recent publications by Koch and colleagues on the "neural correlates of consciousness" (e.g., Rees, Kreiman, & Koch, 2003) still only touch upon cellular physiology, and say nothing about possible underlying molecular mechanisms.

In Bickle (2003) I presented experimental evidence that neurophysiology was discovering cellular mechanisms of some aspects of conscious experience. The work I discussed included single-cell recordings and manipulations in primate dorsolateral prefrontal cortex during "working memory" tasks, changes to action potential frequencies in individual visual neurons with changing focus of selective attention, and sensory experiences induced by cortical microstimulation. Still, even this work remained removed from the molecular wave that washed through neuroscience since the mid-1980s. Molecular neurobiology has now even penetrated the textbooks. Eric Kandel, James Schwartz, and Thomas Jessell open the fourth edition of their monumental *Principles of Neural Science* with the following remark:

> This book . . . describes how neural science is attempting to link molecules to mind – how proteins responsible for the activities of individual nerve cells are related to the complexity of neural processes. Today it is possible to link the molecular dynamics of individual nerve cells to representations of perceptual and motor acts in the brain and to relate these internal mechanisms to observable behavior. (2000, pp. 3–4)

Molecular biology's impact on neuroscience has even generated a new field, "molecular and cellular cognition," whose stated goal is to "derive explanations of cognitive processes that integrate molecular, cellular, and behavioral mechanisms" (www.molcellcog.org/about_us.htm). This field is thriving, with new results being published constantly in major scientific journals. All of this suggests the question: Why not "molecular consciousness studies"?

The first purpose of this chapter is to bring to wider attention already accomplished work on the molecular mechanisms of conscious experience. I will explain one set of recent experiments in detail and speculate about their general promise. My approach – describing the reduction of specific features of consciousness – emphasizes the nature of real reductionism in neuroscientific practice. The painstaking, experimentally intensive techniques of cellular and molecular neuroscience generate "creeping rather than sweeping" reductions, bearing little resemblance to the grandiose examples that historians and philosophers of science take as exemplars.[1] My approach here is *metascience* (Bickle, 2003): examining detailed scientific case studies in an effort to understand the nature of a relationship (like reduction) in actual scientific practice, rather than imposing philosophical intuitions on an analysis of what that relation "has to be." Thus I'll present far more experimental details than philosophers are used to reading. In the final section I'll argue that these details constitute yet another example of the "intervene cellularly/molecularly and track behaviorally" account of reductionism-in-practice I've articulated and defended in recent papers (e.g., Bickle, 2006a, 2006b). My conclusion will be that "ruthless" psychoneural reductionism's assault on consciousness is already underway.

2. Two Phenomenologically Distinguishable Features of Conscious Experience

Conscious experience possesses features that can be distinguished phenomenologically. Two examples are awareness and arousal state. Awareness pertains to consciously monitoring our external and internal environments. Being conscious at the moment, you are aware: visually, of the page in front of you; auditorially, of footsteps in the hallway; kinesthetically, of hunger pangs rumbling through your abdomen; and so on. Were you to doze off into dreamless sleep, you would no longer be consciously aware – of those events or of any others. One common component of our "internal" awareness is a state of general arousal. Perhaps video footage of a recent world calamity has put you on edge. Or perhaps yesterday's late-night escapades have you hung over. While often keyed to determinate events in our internal and external environments, these states typically lack the determinate content of other components of our conscious awareness.[2]

It seems reasonable to hypothesize that these phenomenologically distinguishable features of conscious experience depend upon distinct brain mechanisms. But brains are composed of entities and interactions at many levels of biological organization: systems, circuits, cells, subcellular organelles, and macromolecules – not to mention the behavioral and information-processing levels above these and the general chemical and biophysical levels below. Where in this hierarchy should we expect to find the neural mechanisms producing these distinguishable features of phenomenal consciousness? Recent experimental work suggests a surprising answer: at the macromolecular level of agonistic activities at subunits of γ-amino-butyric acid type A ($GABA_A$) receptor proteins (Jurd et al., 2002; Löw et al., 2000; McKernan et al., 2000; Reynolds et al., 2003). Molecular biological tools for manipulating gene expression and protein synthesis within neurons, down to the level of single amino-acid residues, provide intervention techniques in behaving mammals. Performances in well-accepted behavioral paradigms for quantifying awareness and arousal state provide techniques for tracking specific behavioral effects. These empirical results support the conclusion that awareness and arousal state are linked to agonistic activities involving specific amino acid residues on different subunits of $GABA_A$ receptors.

3. The $GABA_A$ Receptor Protein and a Variety of Genetically Mutated Mice

No doubt the scientific assertions in the last few sentences of the previous paragraph are opaque to many philosophers and cognitive scientists. We thus turn to a brief description of the $GABA_A$ receptor protein (drawing primarily on Chebib & Johnston, 2000). The neurotransmitter molecule GABA influences neurons via three major classes of protein receptors: $GABA_A$, $GABA_B$, and $GABA_C$. $GABA_A$ receptors are components of the major inhibitory neurotransmitter system in the mammalian central nervous system. These receptors are ionotropic ("fast" receptors): when GABA or GABA agonist molecules binds to them, their three-dimensional configuration changes to open a direct channel for chloride ions (Cl^-).[3] Despite the neuron membrane's resting negative potential (intracellular relative to extracellular, around – 55μV), Cl^- ions enter the neuron through these opened receptors owing to a strong concentration gradient. (There are far more Cl^- ions outside the cell membrane compared to inside.) Cl^- influx hyperpolarizes the neuron membrane beyond its resting potential, making that patch of membrane further inhibited to excitatory post-synaptic potentials (EPSPs).

The $GABA_A$ receptor is a pentamer (composed of five protein subunits). Each subunit has a large extracellular N-terminal domain, followed by three membrane-spanning domains (M1-M3), followed by an intracellular loop and a fourth membrane-spanning domain (M4), and ending with an extracellular C-terminal (Figure 1A). Being components of proteins, each subunit is itself composed of amino-acid sequences. The extracellular N terminal domain contains part of the receptor's agonist/antagonist binding site. The five subunits are arranged such that each M2 membrane-spanning domain forms a wall of the Cl^- ion channel pore. The charges of the specific amino acids making up the M2 domains determine the receptor's selectivity for negatively-charged Cl^- ions (Figure 1B). $GABA_A$ receptors are heteromeric, that is, they are composed of different collections of protein subunits (one example of which is illustrated in Figure 1C). With six distinct α, four β, three γ, one δ, and one ε subunits now described, there are over 2,000 possible combinations of $GABA_A$ isoforms containing its characteristic pattern of 2α, 2β, and one other subunit type.[4] However, the number of isoforms estimated actually to exist *in vivo* is less than ten (McKernan & Whiting, 1996, although their estimate is not without published controversy, e.g., Sigel & Kannenberg, 1996; Barnard et al., 1998). The intracellular loop between membrane-spanning domains M3 and M4 contains binding sites for various protein kinases that affect intracellular signaling transduction and membrane-clustering of $GABA_A$ receptors.

$GABA_A$ receptors are a common target for clinically relevant concentrations of numerous anesthetic drugs. Most intravenous anesthetics, like pentobarbital or propofol, lack selectivity for specific GABA receptor classes, much less for specific $GABA_A$ receptor subunits (Belelli et al., 1997). Etomidate is an important exception. It shows agonistic selectivity *in vitro* for $GABA_A$ receptors containing β_2 and β_3 subunits. This selectivity depends solely upon an asperagine residue at position

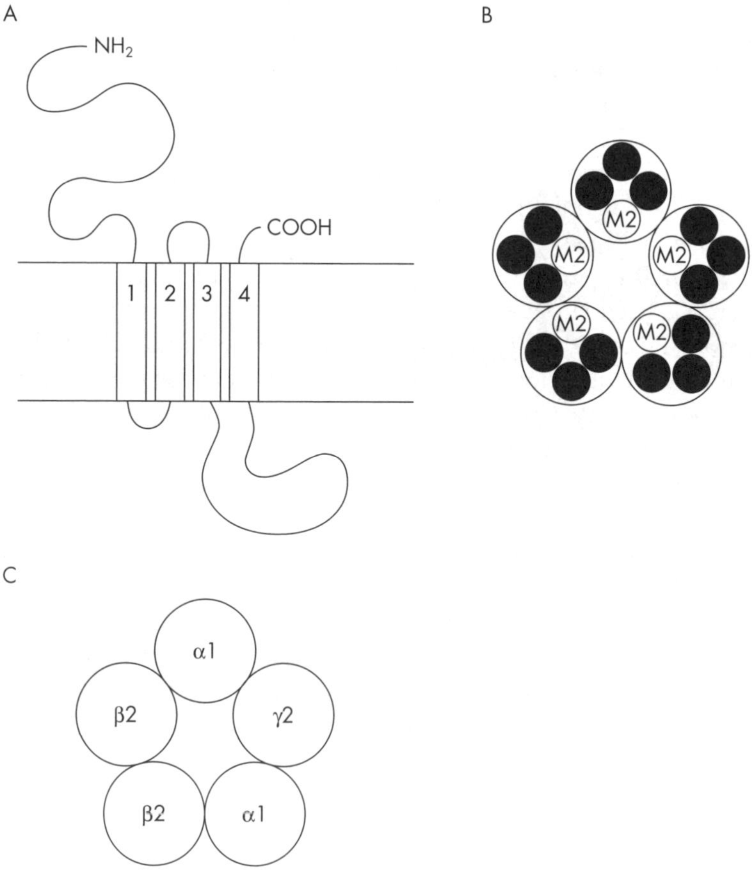

Figure 12.1 Structure of $GABA_A$ receptor protein. **A**. General structure of each subunit, displaying extracellular amine (NH_2) and carboxyl (COOH) terminals, four membrane spanning domains, and intracellular loop. **B**. pentameric arrangement of the subunits, with membrane-spanning domain M2 of each subunit forming the pore of the ion channel. **C**. Components of a $GABA_A$ receptor composed of two α protein subunits, two β protein subunits, and one γ subunit.
Source: Chebib and Johnston (2000, p. 1428). Reprinted with permission from the American Chemical Society. Copyright © 2000 American Chemical Society.

265 on the membrane-spanning M2 domain of both subunits (Belelli et al., 2003).[5] Its general anesthetic action and subunit-specific selectivity make it useful for investigating the molecular mechanisms of features of consciousness *in vivo*.

The idea of using anesthetic agents to study the mechanisms of consciousness is not novel. No less an authority in consciousness studies as anesthesiologist Stuart Hameroff wrote (nearly one decade ago):

> In the past ten years a resurgence of interest in the problem of consciousness has emerged across neuroscience, cognitive science, philosophy, physics, molecular biology, and many other disciplines. The study of general anesthetics lies at a critical juncture and offers a unique and valuable paradigm. Anesthesiology should advance toward, not retreat from the problem of consciousness. (1998, p. 31)

Since many anesthetics are either excitatory synapse antagonists or inhibitory synapse agonists, the basic logic of experimental design is to administer the drug to an experimental group with a targeted receptor mutation and to a "wild-type" control group, and compare performances on well-accepted behavioral measures of awareness and arousal state.[6] Statistically comparable behavior measures across groups indicate that activity at the site of the receptor mutation is not part of the neuromolecular mechanism for that feature of conscious experience. Statistically significant behavioral differences, especially increases in measures of some feature of consciousness in the mutants under drugged conditions, indicate that activity at the site of the receptor mutation is part of the neuromolecular mechanism for that feature. For then the engineered receptor mutation has rendered the drug incapable of exerting its usual anesthetic effects.

Reynolds et al. (2003) used a gene-targeting technique in mouse embryonic stem (ES) cells to replace the asparagine residue at position 265 on the M2 domain of the β_2 subunit with a serine residue. (Serine was chosen because it occupies the equivalent position on the $GABA_A$ β_1 receptor subunit, at which etomidate is inactive.) Standard breeding techniques generated homozygous β2N265S mutants with the single asparagine-to-serine replacement and wild-type littermates. The mutants were phenotypically normal compared to wild types on standard behavioral and clinical screens. (These included measures of spontaneous motor activity during both light and dark conditions and responsiveness to noxious stimuli. Similar controls were used for all genetically engineered mutants described below.) Molecular genetic measures confirmed that the induced mutation did not affect the number of GABA binding sites on β_2 subunits throughout the brains (compared to wild types). Locations and densities of $GABA_A$ receptors containing β_2 subunits in the mutants corresponded well to previous reports in the wild type. Previous studies had revealed that Purkinje neurons in the mouse cerebellum express mostly β_2-containing $GABA_A$ receptors. Reynolds et al. (2003) showed that these cells in the mutants had lost all sensitivity *in vitro* to etomidate. However, GABA application (whose activity is not dependent on the asparagine residue) induced identical inhibitory currents in wild-type and β2N265S mutant cerebellar Purkinje neurons. But co-application of a subthreshold quantity of GABA and etomidate potentiated inhibitory currents only in wild-type neurons. In contrast, potentiation of GABA induced currents in both wild-type and mutant neurons was statistically similar with coapplication of subthreshold quantities of GABA and the barbiturate pentobarbital (which, to repeat, is not selective for β_2-containing $GABA_A$ receptors). These phenotypic screens and *in vitro* investigations clearly indicate that these β2N265S mutant mice had the asparagine-to-serine mutation correctly placed, with the predicted physiological effects.

Using similar ES cell gene-targeting techniques, Jurd et al. (2002) replaced the asparagine residue at position 265 on the M2 domain of $GABA_A$ β_3 subunits with a methionine residue. As with the β2N265S mutants, these β3N265M mutants also showed no phenotypic abnormalities in standard screens. All major $GABA_A$ subunits, including β_3, were expressed and distributed across the brains at normal levels. The single-point mutation had no effect *in vitro* on GABA (or benzodiazapine) binding and inhibitory current. In electrophysiological studies similar to those just described for the β2N265S mutants, etomidate binding *in vitro* and potentiation of GABA induced inhibitory currents were significantly diminished compared to wild-type controls. These phenotypic screens and *in vitro* investigations clearly indicated that these β3N265M mutant mice had the asparagine-to-methionine mutation correctly placed, with the predicted physiological effects.

Having established the correctly placed and stable insertion of these single-point mutations, *in vivo* studies could now track any behavioral effects of these molecular interventions using standard experimental methods for assessing features of consciousness. The results obtained so far constitute early steps toward mind-to-molecular activities reductions, at least in the way that reduction works in practice in current molecular and cellular cognition.[7]

4. Behavioral Measures of Rodent Awareness

What tests using rodents provide behavioral measures for features of phenomenal consciousness? One must be willing to "operationalize" these features if one is to address them experimentally. This "operationalism" is strictly methodological, *for the purpose of experimental investigation only*. It is *not* intended as conceptual or semantic analysis. To deny this possibility is simply to rule consciousness out of science's court – not just molecular biology's. Some fans of the phenomenal features of conscious experience seem willing to do this, but it strikes me that most investigators in our present "naturalistic milieu" at least hold open the possibility of consciousness being within science's reach. This chapter is only addressed to the latter.

Of course, the behavioral tests we accept as measures of these features should cohere with our phenomenology. Choices of behavioral measures for awareness and arousal state can be judged to be poor ones if they stray too far from our experiences of these features. One can challenge even popular behavioral measures on grounds that they stray too far. But such challenges must criticize the specific tests employed, and so one must first know about the tests which actually are employed. One might also reasonably ask critics for help in developing new and better measures. Naysaying is easy. Pushing serious science forward is not.

In light of these remarks, a word is in order on the use of rodent models for studying the molecular mechanisms of consciousness. Even if successful, why are experiments with these mutated mice revealing of anything more than possible mechanisms of rodent awareness and arousal? First, the biochemistry and physiology of $GABA_A$ receptors are virtually identical across mammalian species. Second, the anesthetics employed exert similar effects across species, both *in vitro* and *in vivo*. Third, as we will see below, some of the behavioral measures used to assess rodent awareness and arousal state experimentally have analogs used in human clinical anesthesiology. I suppose that these reasons won't convince the persistent skeptic who refuses to ascribe phenomenal consciousness to anything but humans (or him- or herself). Fortunately there aren't too many such skeptics and the appropriate response is to point out the scientific idleness of their view.

In the rodent experimental literature, three tests are popular for assessing awareness: loss of righting reflex, loss of pedal withdrawal reflex, and onset and maintenance of "burst suppression" in the electroencephalograph (EEG). Reynolds et al. (2003) used all three tests, comparing performances of wild-type mice and β2N265S mutants under conditions of anesthetic dosages. Jurd et al. (2002) used the first two tests, with wild types β3N265M mutants.

Loss of Righting Reflex (LORR). Following anesthetic administration, animals first are observed until they lose the ability to regain normal posture after slumping. They are then laid on their backs and measured for the amount of time it takes them to regain normal posture (LORR duration). This test is used as a measure of loss of awareness and so of the hypnotic properties of an anesthetic. The amount of time spent on its back provides a quantified measure of the animal's loss of awareness due to the anesthetic.

Reynolds et al. (2003) compared LORR durations in β2N265S mutant mice and wild-type littermates under different anesthetic conditions. Under low (20 mg/kg of body weight, administered via intraperitoneal [i.p.] injections), medium (30 mg/kg i.p.), and high (40 mg/kg i.p.) doses of etomidate, LORR in both wild types and mutants were dose dependent (longer LORR durations obtained with increasingly higher dosages) and statistically similar (Reynolds et al., 2003, Figure 4A). LORR duration in wild types and β2N265S mutants were also statistically similar following i.p. injections of pentobarbital, the hypnotic barbiturate that is not selective for β_2 subunits (or for $GABA_A$ receptors). These results show that loss of awareness induced by etomidate, and so the drug's hypnotic properties, are mediated independently of agonistic activity at $GABA_A$ β_2 receptor subunits (since the β2N265S mutants lack this activity). They suggest that the neural mechanisms of phenomenal awareness do not include inhibition of GABA activity at receptors containing the β_2 subunit.[8]

Jurd et al. (2002) compared wild types and β3N265M mutant mice on LORR using a variety of anesthetics. Under low (5 mg/kg i.v.), moderate (10 mg/kg i.v.), and high (15 mg/kg i.v.) etomidate dosages, both groups demonstrated a dose-dependent response, but these durations were reduced in the β3N265M mutants by greater than 50 percent (statistically significant, $p < 0.01$). For example, at the moderate (10 mg/kg) dosage, wild types displayed LORR for nearly 40 minutes; β3N265M mutants averaged just slightly over 10 minutes (Jurd et al., 2002, Figure 3). The high etomidate dosage (15 mg/kg, i.v.) was lethal for nearly half of wild types but for none of the β3N265M mutants. Jurd et al. (2002) obtained similar LORR duration differences and death rates between wild types and β3N265M mutants using the anesthetic propofol, a subunit nonselective GABA agonist. However, the steroid anesthetic combination of alphaxolone and alphadolone produced statistically similar LORR durations and death rates at high dosages across the two groups. This important control result indicates that the large alterations of sensitivity to $GABA_A$ receptor agonists in β3N265M mutants were unlikely to have resulted from changes in neuronal circuitry or other gross abnormalities (which would have rendered them insensitive to steroid anesthetics as well).

Loss of Pedal (Hind Limb) Withdrawal Reflex (LOPWR). LOPWR is a common measure of surgical anesthesia in experimental animals. Following anesthetic administration, rodents are laid on their backs with a hindlimb extended slightly. Atraumatic forceps are used to pinch the webbing between the digits of the extended limb. A clear attempt to withdraw the limb is recorded as regaining the PWR. Typically LOPWR is tested in alternate limbs every 1–3 minutes until PWR is regained. The amount of time after anesthetic dosage until the PWR is regained is a quantitative measure of loss of conscious awareness and of the hypnotic properties of an anesthetic.

Following both etomidate and propofol administration, Reynolds et al. (2003) found dose-dependent increases in LOPWR duration in both wild-types and β2N265S mutant mice (although duration was somewhat shorter for mutants at high i.v. dosages; see note 8). This is consistent with other measures of awareness following etomidate anesthesia in the β2N265S mutants. On the other hand, Jurd et al. (2002) found significant differences in LOPWR duration between wild types and β3N265M mutant mice following low, moderate, and high i.v. etomidate and propofol dosages, but no significant differences following steroid anesthetic dosages (combined alphaxolone/alphadolone). Even 15 mg/kg i.v. of etomidate or propofol failed to induce LOPWR for more than one minute in the β3N265M mutants; these doses were lethal to more than half of wild types and induced LOPWR in surviving wild types for 10–15 minutes. However, at high dosages of the steroid anesthetic, β3N265M mutants and wild types had statistically comparable LOPWR durations for up to 10 minutes. As in the LORR studies just discussed, this control suggests that the highly significant differences between wild types and β3N265M mutants under etomidate and propofol anesthesia are due to the single-point mutation on β_3-containing $GABA_A$ receptors, rather than to a change in neural circuitry.

Anesthesia measures using EEG. Many anesthetics (including etomidate) produce a "burst suppression" EEG pattern characterized by large-amplitude fast spikes interspersed between relatively low-amplitude, low-frequency signals. This pattern differs significantly from both the low-amplitude, high-frequency signals associated with wakeful awareness and the higher amplitude, lower-frequency patterns of slow-wave sleep (Reynolds et al., 2003, Figure 6). Interestingly, burst suppression is a measure of anesthesia used in human surgery, with EEG measured from scalp electrodes.

Reynolds et al. (2003) implanted both wild types and β2N265S mutant mice with radiotelemetry transmitters for recording EEG activity. In the anesthesia EEG experiments, etomidate was administered intravenously at moderate (10 mg/kg i.v.) and high (12.5 mg/kg i.v.) dosages. EEG patterns were analyzed in 100 millisecond (msec) epochs for percentages of burst suppression versus burst activity per minute. The percentage of time spent in burst suppression was dose dependent for both wild types and β2N265S mutants (with percentage of time in burst suppression significantly higher at higher dosages.) There were no statistically significant differences between wild-type and β2N265S mutants receiving the same dosages.

Results from these three common measures of loss of awareness to anesthetic administration on β2N265S and β3N265M mutant mice and their wild-type littermate controls show that the anesthetic (loss of awareness) effects of β_2/β_3 selective etomidate is mediated through its agonistic activity at β_3 subunits alone, independent of activity at β_2 subunits (mutated in the unaffected β2N265S mice). Inhibition of GABA and GABA-agonist activity at the β_2 subunit of $GABA_A$ receptors appears not to be a component of the molecular mechanisms producing conscious awareness; but inhibition of that activity at the β_3 subunit appears to be a component.

5. Behavioral Measures of Rodent Arousal State

In the rodent experimental literature, three tests are popular for assessing arousal state: spontaneous locomotor activity, trained rotorod performance, and percentage of time spent in slow-wave sleep following anesthesia recovery. Reynolds et al. (2003) used all three tests, comparing performances of wild types and β2N265S mutants under conditions of etomidate and nonselective anesthetic dosages.[9]

Spontaneous locomotor activity. One test of a drug's sedative effects on arousal state involves delivering a subthreshold dose insufficient to produce loss of awareness and observing the animals' subsequent movements. Less spontaneous locomotor activity reflects higher sedation due to the drug and hence a lower arousal state. Reynolds et al. (2003) administered doses of etomidate ranging from 0.3 to 12.5 mg/kg i.p. to β2N265S mutant mice and wild-type littermates, then placed the animals in an activity chamber fitted with intersecting light beams. Under normal (non-sedated) conditions, mice readily explore the new environment. They break the light beams many times, indicating many cage crossings, high exploratory activity, and a high arousal state.

Not surprisingly, the higher doses of etomidate (sufficient to induce a brief LORR) produced low levels of activity in both groups (measured by few beam crossings over a 20-minute period). However, lower (subthreshold) doses (between 3 to 7.5 mg/kg i.p.) had a dose-dependent effect on wild types – higher doses led to fewer beam crossings – but had *no* effect on β2N265S mutants. The mutants had significantly ($p < 0.05$) greater activation levels compared to wild types for all dosages administered in the subthreshold sedation range. This result indicates that agonistic activity at $GABA_A$ receptor β_2 subunits mediates the sedative properties of etomidate, and therefore the animal's arousal state.

Recovery from anesthesia measured by trained rotorod performance. The rotorod has a diameter that rodents can perch upon. It revolves at a constant speed. Animals are trained at a particular speed until they can maintain balance for a specified time on some number of consecutive trials. Following recovery of awareness from anesthesia, the time that a trained animal can balance on the rotorod is a measure of the drug's sedative effects, and so of the animal's changing arousal state. The less time that a trained animal can remain balanced indicates a higher level of maintained sedation, and hence a lower arousal state.

Reynolds et al. (2003) trained wild types and β2N265S mutant mice on a rotorod (revolving at 16 rpm) until each animal could remain balanced for 120 seconds on three consecutive trials. Trained mice were then administered low (5 mg/kg), moderate (10 mg/kg), or high (15 mg/kg) doses of etomidate i.v. or a low dose (30 mg/kg) of propofol i.v. All dosages were sufficient to induce LORR as a measure of loss of awareness. Animals were then tested on the rotorod at various times after they had recovered their righting reflexes. As expected, there were no statistically significant differences in recovery of rotorod performance between wild types and β2N265S mutants at all times following recovery from the GABA non-selective anesthetic propofol. However, at all doses of etomidate, β2N265S mutants quickly regained trained rotorod performance levels: within 3 minutes of regaining righting reflex for both 5 and 10 mg/kg dosages and within 9 minutes for 15 mg/kg. Wild types, on the other hand, only regained trained rotorod performance levels within 27 minutes after regaining righting reflex for the 5 mg/kg etomidate dosage and never regained trained levels for the entire 30 minutes measured for 10 and 15 mg/kg doses. These differences were statistically significant ($p < 0.0005$) between wild-type and β2N265S mutants for all dosage groups. These results again indicate that agonist activity at $GABA_A$ β_2 subunits mediate the sedative effects of etomidate, suggesting its role in the molecular mechanisms producing conscious arousal state.

Sedation measures using EEG. Changes in the amount of time that an animal spends in slow-wave sleep, characterized by an EEG pattern of higher-amplitude, lower-frequency waves, indicates its changing level of sedation. The more time the animal spends in slow-wave sleep, the less aroused it is. Using anesthetic dosages, individual animals are first scored for baseline amounts of slow-wave sleep at various times on day 1 (the day prior to anesthetic administration). On day 2, after recovery of awareness from anesthetic administration, they are assessed for amounts of slow-wave sleep at various times, with measures assessed as a percentage compared to baseline (matched for circadian pattern).

Reynolds et al. (2003) conducted a sedation EEG study using the same wild types and β2N265S mutant mice they had used for the anesthetic EEG studies discussed in the previous section. Recall that in the anesthetic EEG studies, they found no significant differences across the two groups for all dosages of etomidate. In the sedation studies, however, they found statistically significant differences in recovery time across the two groups at both moderate (10 mg/kg i.v.) and high (12.5 mg/kg i.v.) etomidate doses. The wild types administered the moderate dose spent more than twice the amount of time in slow-wave sleep at 1 and 2 hours following recovery compared to their baseline rates from the previous day, while β2N265S mutants were statistically similar to their baseline rates at both times. Wild types administered the higher dosage spent over 400 percent more time in slow-wave sleep 3 hours following recovery (again, compared to their own baseline rates), while the β2N265S mutants remained statistically similar to their baseline rates. So while the anesthetic EEG studies in these rats indicated that agonistic activity at $GABA_A$ β_2 subunits is not part of the molecular mechanisms for awareness, the sedation EEG studies *in these same mice* showed that it does mediate level of sedation following anesthetic recovery, and so is a part of the molecular mechanisms for arousal state.

Taken together, Reynold et al.'s (2003) and Jurd et al.'s (2003) results with $GABA_A$ receptor β subunit mutant mice show that hypnotic anesthesia and sedation induced by GABA agonistic anesthetics are mediated by activity at distinct receptor subunits. Given the connection between hypnotic anesthesia and conscious awareness, and between sedation and conscious arousal state, we can surmise that while activity at $GABA_A$ β_2 subunits is not part of the molecular mechanisms for awareness, it is part of the mechanisms for arousal state. Contrariwise, activity at β_3 subunits seems to be part of the molecular mechanisms for awareness. Experimental double dissociations like these are a recognized method for "localizing" functions to particular neural regions or mechanisms.[10]

6. Reduction-in-practice in Molecular and Cellular Consciousness Studies

These molecular-genetic and pharmacological interventions and behavioral results are novel and ingenious. But it is now time to take on the question of whether they constitute steps toward an eventual *reduction* of features of conscious experience to activities at specific $GABA_A$ receptor subunits in individual neurons. Most philosophers and cognitive scientists will find such a reductionist claim baffling, even if they aren't predisposed against the relevance of neuroscience for explaining cognition. Work on intertheoretic reduction since 2000 bears out their worry.[11] Whether the philosophical concerns motivating these accounts are epistemological (e.g., achieving deductive or explanatory unification across theories), metaphysical (e.g., achieving ontological unification via cross-theoretic object, property, or event identities) or historical (e.g., accounting for a handful of "achieved" reductions, mostly from the history of physics), philosophical accounts of reduction have shared a "step-wise" picture, without any notable exceptions.

This picture attitude goes back explicitly to (at least) Paul Oppenheim and Hilary Putnam's (1958) classic treatise on "the unity of science." They were explicit about the role of "microreductions" in achieving unified science: "The only method of attaining unitary science that appears to be seriously available at present is micro-reduction" (1958, p. 6). They clearly specified the levels and their arrangements in science's hierarchy:

6. .... Social groups
5. .... Multicellular living organisms
4. .... Cells
3. .... Molecules
2. .... Atoms
1. .... Elementary particles (1958, p. 9)

And they were univocal about the steps necessary to bridge the descending levels: "We maintain that each of our levels is *necessary* in the sense that it would be utopian to suppose that one might reduce all of the major theories or a whole branch concerned with any one of our six levels to a theory concerned with a lower level, *skipping* entirely the *immediately* lower level" (1958, p. 10; original emphasis).

This step-wise picture is one of the few significant features of Oppenheim and Putnam's account that has survived subsequent work. It remains a part of every current reductionist program. Even Patricia Churchland, whose reductionist credentials are rarely assailed, has recently written that "the idea of explanation in a single bound" does stretch credulity, *and neuroscientists are not remotely tempted by it*" (2002, p. 29; my emphasis). Indeed, more recent and detailed accounts of the mind–brain sciences (since Oppenheim and Putnam's) have inserted more levels to move through, one step at a time, between psychology and molecular neuroscience. There is the domain of cognitive science in general, which treats cognitive systems as information processors. There are the domains of cognitive and computational neuroscience: neural systems, regions, and circuits. There is neuroanatomy, from the gross down to the micro. There is cellular physiology, studying the activities of individual neurons. The reductionist's road seems a long one.

So if reduction is the relationship being developed in the "mind-to-molecules" linkages being established experimentally in the work I discussed above, isn't the step-wise procedure being violated? Isn't this work seeking to reduce features of consciousness directly to molecular activities, bypassing the intermediate levels entirely? Is this what psychoneural reduction has become?

Let's begin to address these questions with a metaphilosophical point. Suppose that we want a realistic picture of what "scientific reduction" is. I contend that we'd best achieve this goal by looking closely at the practices in the best available examples from scientific disciplines that seek lower-level mechanistic explanations of higher-level phenomena.[12] Molecular and cellular cognition is just such a "reductionistic" science. Its practitioners label themselves so, and are so labeled by their colleagues across the neurosciences. And publication of experimental results in journals like *Journal of Neuroscience* and *The FASEB Journal* – along with *Cell*, *Neuron*, *Science*, *Nature*, *Nature Neuroscience*, *PNAS*, and others – is one accepted mark of significant research by scientists in this field. Hence it is to such cases we should turn. Once we discover in them what "reductionism-in-practice" is, we can then look back to see how well the real relationship fits with epistemic, metaphysical, and historical ideals that motivated philosophers' attention.

What, then, is reductionism-in-practice in molecular and cellular cognition? As I've urged previously, it involves experimental results that meet a two-part methodology:

- *Intervene causally* into cellular activity or molecular pathways within specific neurons in behaving animals (e.g., via genetically engineered mutations);
- *Track* statistically significant differences in the behavior of these animals using behavioral protocols and measures well accepted within experimental psychology for the cognitive phenomenon under investigation.

This methodology constitutes a necessary condition on experimental success in this acknowledged reductionistic field. Molecular and cellular cognitivists only claim a successful *mechanistic explanation* or *reduction* after they have intervened successfully into that proposed mechanism *in vivo* and tracked statistically significant behavioral effects – after they have engineered a higher-level effect, as it were, via lower-level interventions (Bickle, 2003, 2006a, 2006b).

Of course there are other experimental methodologies employed in this field's research. Much "intervene behaviorally and track behaviorally" research must first be accomplished by experimental psychologists before we have appropriate behavioral measures for specific cognitive phenomena. Much "intervene behaviorally and track cellularly/molecularly" must be done, by cognitive and behavioral neuroscientists on down to neurophysiologists, to learn where to start inserting our cellular and molecular interventions. Much "intervene cellularly/molecularly and track cellularly/molecularly" must be done by basic neurobiologists (and molecular biologists in general) to both learn where to insert the interventions (even more precisely) and to perfect the intervening techniques. But from the perspective of molecular and cellular cognition, all of this additional work is *essentially heuristic* for the key kind of experimental results that ultimately establish the lower-level mechanisms – the *reductions* – of phenomena described using higher-level resources. When the "intervene molecularly and track behaviorally" experimental results are finally in, against a theoretical background informed by the kinds of results just surveyed, the lower-level mechanisms, in conjunction with the anatomical circuitries containing the neurons intervened into, are taken to *directly explain* the behavioral data. These "final" explanations *set aside causal–mechanistic explanations* offered at intermediate levels of theorizing, including the psychological, the cognitive scientific, and the cognitive neuroscientific.

Taken at face value, the results surveyed in sections 4 and 5 above fit the "intervene molecularly and track behaviorally" model of reductionism-in-practice that I've gleaned from molecular and cellular cognition generally. This "fit" is ultimately the best argument that this work constitutes a start toward a reduction-in-scientific-practice of features of consciousness. It is the beginning of a related field, "molecular and cellular consciousness studies." In all those cases, the key experimental results involved bioengineered genetic mutations to *intervene* at a single amino-acid residue in a subunit of the $GABA_A$ receptor protein and various pharmacological agents known to interact with those receptors either at that subunit or some other.

The experimenters then *tracked* the behavioral effects of these interventions using behavioral measures of awareness and arousal state prominent in rodent experimental work.

As a further consideration, consider the discussions of the scientists achieving these results themselves. They don't speak the language of philosophers, especially in their primary experimental reports, so one doesn't always find explicit claims to "reduction." Still, their claims are suggestive. Summarizing their findings with β2N265S mutant mice in the paper's abstract, Reynolds et al. write:

> Loss of pedal withdrawal reflex and burst suppression in the electroencephalogram were still observed in the mutant mice, indicating that loss of consciousness can be *mediated purely* through β_3-containing receptors. The sedation produced by subthreshold doses of etomidate and propofol during recovery from anesthesia was present only in wild-type mice, indicating that the β_2 subunit mediates the sedative properties of anesthetics. These findings show that anesthesia and sedation are mediated by distinct $GABA_A$ receptor subtypes. (2003, p. 8608; my emphasis)

They indicate clearly the direct link that already exists between clinical anesthesia and conscious awareness operationalized using the behavioral measures employed: "Our understanding of what happens during clinical anesthesia is still rudimentary; however, it is clear that the major requirements are unconsciousness as determined by lack of sensitivity to noxious stimuli, amnesia and preferably lack of movement" (2003, p. 8616). And they speak to the direct explanatory/reductive potential of intervening molecularly and tracking behaviorally against a background of neuroanatomical circuitries: "Using a combination of genetically modified mice and receptor subtype-specific compounds, we will be able to *more clearly define* the mechanisms of actions of anesthetics and our understanding of which neuronal pathways are involved. It may also enable a closer understanding *of what differentiates* sleep and unconsciousness" (2003, p. 8616; my emphasis).

Jurd et al. offer similar assessments.[13] In their abstract they summarize their conclusion: "Taken together, our results demonstrate that a single molecular target, and indeed a specific residue (N265) located within the $GABA_A$ receptor β3 subunit, is *a major determinant* of behavioral responses evoked by the intravenous anesthetics etomidate and propofol" (2002; my emphasis). Commenting on their results, they write that "the striking abolition of the immobilizing response to etomidate and propofol that we observe in the β3N265M mice reveals for the first time *an essential role* of the β3 subunit in *the mediation* of this endpoint of anesthetic action" (2002; my emphasis) – namely, conscious awareness operationized behaviorally. They conclude by claiming to have shown "for the first time *in vivo* that a specific residue located within the second transmembrane region of the β3 subunit of the $GABA_A$ receptor has *a profound influence* on behavioral general anesthetic responses" (2002; my emphasis). This isn't terminology favored by philosophers; but coupled with the "fit" stressed above between their experimental practices and those of molecular and cellular cognitivists generally, whose work forms the core of the "ruthless reductionist's" metascientific account, these scientists' claims are plausibly interpreted as reductionist.

The implications of this recent experimental work are clear: molecular and cellular consciousness studies are following in the "ruthlessly reductive" footsteps of molecular and cellular cognition more generally. Who says we can't do a molecular biology of consciousness? Nobody should, in light of research programs doing it as you read this – and progressing quite nicely.

Notes

1 "Creeping rather than sweeping" is a pithy phrase I learned from Kenneth Schaffner (unpublished correspondence).
2 Anxiety level is a third phenomenologically distinguishable feature of conscious experience, and is also one for which experiments similar to the ones described below have already been performed. Limitations of space preclude discussion of these further experiments here, although I'll mention some of this work in subsequent notes.
3 An agonist *enhances* neurotransmitter activity at synapses; an antagonist *inhibits* neurotransmitter activity. Since GABA itself is an *inhibitory* neurotransmitter, a GABA agonist (like the ones about to be discussed) enhances its inhibitory activity.
4 There are two splice variants of both β and γ subunits.
5 Asparagine and serine are particular amino acids, as is methionine (discussed below).

6 "Wild types" are conspecific animals without the engineered mutation, often littermates of the mutants. They provide important experimental controls in the kinds of experiments to be described.

7 These two research groups have also engineered point mutations at single amino-acid residue in $GABA_A$ α subunits, with similar successes. These mutants were then used, along with their littermate wild-type controls and benzodiazepine anxiolytic (anxiety-reducing) drugs, to investigate the molecular mechanisms of anxiety states using a number of behavioral tests popular for rodent models.

8 Reynolds et al. (2003) found statistically significant differences in LORR and Loss of Pedal Withdrawal Reflex duration (described below) between wild types and β2N265S mutants with moderate to large dosages of etomidate delivered via intravenous (i.v.) tail injections. They attribute this result not to the missing $GABA_A$ β_2 subunits, but rather to different pharmacodynamics of etomidate, depending on route of administration. Etomidate absorbs much faster with i.v. versus i.p. injections, probably resulting in differential occupation of $GABA_A$ β_3 receptor subunits in wild types and β2N265S mice (since the latter lack functional $GABA_A$ β_2 subunits for etomidate to bind to). Results from Jurd et al.'s (2002) experiments with β3N265M mutants, about to be discussed, are consistent with this interpretation. However, Reynolds et al. admit that more detailed pharmacokinetic and pharmacodynamic studies must be done to clarify this point (2003, 1615). This is one example of additional work required to further develop this potential reduction-in-practice.

9 Both research groups using the β-subunit mutants have also used these measures with α-subunit mutants and benzodiazepine (anxiety-relieving) drugs to dissociate the molecular mechanisms of anxiety level (and anxiolytic properties of these drugs) from those of arousal state (and sedative properties of these drugs). The details are important for establishing the complete current experimental case for a molecular biology of consciousness, but space precludes discussion here.

10 Similar behavioral dissociations using $GABA_A$ receptor α subunit mutants and benzodiazepine anxiolytic drugs (like diazepam, known commercially as Valium), suggest that agonistic activity at α_1 subunits mediate the sedative effects of these drugs (McKernan et al., 2000), while agonistic activity at α_2 subunits mediates the anxiolytic effects (Löw et al., 2000). In keeping with the methodological assumptions of the studies just described, these further results suggest that activities at α_1 subunits are a component of the molecular mechanisms for arousal state, while activities at α_2 subunits are a component of the molecular mechanisms for anxiety level. These studies used the same measures for arousal state described in this section, and a variety of standard behavioral measures

for rodent anxiety (e.g., light–dark choice test, fear-potentiated startle response, and elevated plus maze).

11 See Bickle (2003, ch. 1) for an overview of twentieth-century philosophical work on intertheoretic reduction, with numerous references to the primary literature.

12 So as not to beg any questions, let me simply assert that at the moment questions about the relationship between "ruthless reductionism" and the "new mechanistic" philosophy of science remain open and in need of investigation. One clear difference seems to be the status of multilevel mechanisms. See Bickle (2006a) for a start toward investigating this difference.

13 The quotes that follow are taken from the full text of the article, published online December 3, 2002 (and fully cited in the References below).

References

Baars, B. J. (1988). *A Cognitive Theory of Consciousness.* London: Cambridge University Press.

Barnard, E. A., Skolnick, P., Olsen, R. W., Möhler, H., Sieghart, W., Biggio, G., et al. (1998). International union of pharmacology–XV–subtypes of γ-aminobutryic $acid_A$ receptors – Classification on the basis of subunit structure and receptor function. *Pharmacology Reviews, 50*, 291–313.

Belelli, D., Lambert, J. J., Peters, J. A., Wafford, K. A., & Whiting, P. J. (1997). The interaction of the general anesthetic etomidate with the Gamma-Aminobutryic Acid Type A receptor is influenced by a single amino acid. *Proceedings of the National Academy of Sciences USA, 94*, 11031–11036.

Belelli, D., Muntoni, A.-L., Merrywest, S. D., Gentet, L. J., Casula, A., Callachan, H., et al. (2003). The *in vitro* and *in vivo* enantioselectivity of etomidate implicates the $GABA_A$ receptor in general anesthesia. *Neuropharmacology, 45*, 57–71.

Bickle, J. (2003). *Philosophy and Neuroscience: A Ruthlessly Reductive Account.* Dordrecht: Kluwer Academic Publishers.

Bickle, J. (2006a). Reducing mind to molecular pathways: Explicating the reductionism implicit in "'molecular and cellular cognition." *Synthese*, forthcoming.

Bickle, J. (2006b). Ruthless reductionism in recent neuroscience. *IEEE Transactions on Systems, Man, and Cybernetics*, forthcoming.

Chebib, M. & Johnston, G. A. R. (2000). GABA-activated ligand gated ion channels: Medicinal chemistry and molecular biology. *Journal of Medicinal Chemistry, 43*, 1427–1447.

Churchland, P. M. (1995). *The Engine of Reason, the Seat of the Soul: A Philosophical Journey into the Brain*. Cambridge, MA: MIT Press.

Churchland, P. S. (2002). *Brain-Wise: Studies in Neurophilosophy*. Cambridge, MA: MIT Press.

Crick, F. & Koch, C. (1990). Towards a neurobiological theory of consciousness. *Seminars in the Neurosciences*, *2*, 263–275.

Damasio, A. R. (1999). *The Feeling of What Happens: Body and Emotion in the Making of Consciousness*. New York: Harcourt Brace.

Hameroff, S. (1998). Anesthesia, consciousness, and hydrophobic pockets: A unitary quantum hypothesis of anesthetic action. *Toxicology Letters*, *100/101*, 31–39.

Jurd, R., Arras, M., Lambert, S., Drexler, B., Siegwart, R., Crestani, F., et al. (2002). General anesthetic actions *in vivo* strongly attenuated by a point mutation in the $GABA_A$ receptor β_3 subunit. *FASEB J* (December 3) 10 1096/fj.02-061 1 fje.

Kandel, E. R., Schwartz, J. L., & Jessell, T. (Eds.) (2000). *Principles of Neural Science*. New York: McGraw-Hill.

Löw, K., Crestani, F., Keist, R., Benke, D., Brünig, I., Benson, J. A., et al. (2000). Molecular and neuronal substrates for the selective attenuation of anxiety. *Science*, *290*, 131–134.

McGinn, C. (1989). Can we solve the mind–body problem? *Mind*, *98*, 349–366.

McKernan, R. M., Rosahl, T. W., Reynolds, D. S., Sur, C., Wafford, K. A., Atack, J. R., et al. (2000). Sedative but not axiolytic properties of benzodiazepines are mediated by the $GABA_A$ receptor α_1 subtype. *Nature Neuroscience*, *3/6*, 587–592.

McKernan, R. M., & Whiting, P. J. (1996). Which GABA-A receptor subtypes really occur in the brain? *Trends in Neurosciences*, *19*, 139–143.

Oppenheim, P., & Putnam, H. (1958). Unity of science as a working hypothesis. In H. Feigl, M. Scriven, & G. Maxwell (Eds.), *Minnesota Studies in Philosophy of Science* (Vol. II, pp. 3–36). Minneapolis: University of Minnesota Press.

Ramachandran, V. S., & Blakeslee, S. (1998). *Phantoms in the Brain*. New York: William Morrow.

Rees, G., Kreiman, G., & Koch, C. (2003). Neural correlates of consciousness in humans. *Nature Reviews Neuroscience*, *3*, 261–270.

Reynolds, D. S., Rosahl, T. W., Cirone, J., O'Meara, G. F., Haythornthwaite, A., Newman, R. J., et al. (2003). Sedation and anesthesia mediated by distinct $GABA_A$ receptor isoforms. *Journal of Neuroscience*, *23*, 8608–8617.

Sigel, E., & Kannenberg, K. (1996). $GABA_A$-receptor subtypes. *Trends in Neurosciences*, *19*, 386.

13

MIND READING AND MIRROR NEURONS: EXPLORING REDUCTION

Huib Looren de Jong and Maurice Schouten

1. Introduction: Mind Reading and Reductionism

The topic of this chapter is the possibility of reducing mental states and processes to neural states and processes. As a case study, we focus on what has traditionally been considered a mark of the mental: attributing beliefs and desires to other agents, reading the goals and knowledge behind their behavior. Empathy, folk psychology, mind reading, and shared intention are seemingly simple and commonplace skills. Nevertheless, quite a tangled cluster of issues, spanning philosophy and psychology, is involved. In the philosophy of mind the debate on folk psychology focused on the status of intentional states of beliefs and desires vis-à-vis neuroscience. A traditional philosophical issue is how we can know other minds, and where the concepts for mental states come from: how do we know what beliefs and desires are, and that other people have them? In developmental psychology, the issue is where this skill comes from, and what its psychological mechanisms are. In evolutionary biology, the question is when in evolution empathy emerges, and in anthropology, how universal the belief-desire explanation is across cultures.

An interesting recent development in neuroscience with potentially major relevance for the folk psychology debate is the discovery of so-called mirror neurons in the prefrontal cortex that seem to code for the intention of a movement by an agent. The connection between mind reading and mirror neurons looks like a philosophically rich case since it touches upon the biological basis of social cognition, intentionality, and the biological preconditions of culture. Reductionists should be happy to reconstruct mind reading in terms of neural processes; for example John Bickle (2006) has proposed reducing rat social cognition to a molecular mechanism (but, to be precise, not to mirror neurons). Below, we will defend an antireductionist view of the biological basis of social cognition. We will argue that the reductionist ideal of a straight arrow from mind and social cognition down to (neural or even molecular) neuroscience fails here: rather, the study of the biology of mind reading is suggestive of a patchwork of theories at multiple levels answering different explanatory questions in different contexts. Also, we will show how the new empirical findings suggest revision of the traditional philosophical construal of folk psychology as attributing discrete beliefs and desires to intentional agents. Furthermore, a connection between hermeneutics as the mainstay of the classical humanities and mind reading will be explored, calling into question the traditional chasm between biology and culture.

This will lead to a more general (negative) answer to the question of whether the reduction of understanding to neurons is feasible, i.e., whether ultimately an equipotent image (Churchland, 1989) of the mind-reading capacity can plausibly be constructed bottom-up from the neural theory alone (cf. Bickle, 2006).

2. Folk Psychology and Reduction: Naturalizing Beliefs and Desires?

To set the scene, we start with a brief review of the most important issues in the folk psychology debate in the philosophy of mind. As mentioned, folk psychology was a hot topic in the 1980s and 1990s; the debate focused on the naturalization of folk psychology. Two proposals on how to integrate the belief-desire discourse in the natural world dominated the scene: the computational and the eliminativist.

In the classical computational framework, Fodor (1981, 1990, 1995) argues that beliefs and desires as used in everyday idiom can be identified with propositional attitudes in a Language of Thought, and thus vindicated within a computational theory of mind, more or less in the same way as common-sense physics is vindicated by Newtonian mechanics. Churchland (1981, 1991), in contrast, thinks that beliefs and desires are obsolete, like concepts such as phlogiston or ether; folk psychological discourse in this view has lost contact with developments in the (neuro)sciences, fails to address important areas of mental functioning, and in general is a degenerating paradigm. Hence, its elimination and replacement by the terminology of the neurosciences (neurons, connections, activation vectors, rather than representations, knowledge, judgment) is needed. Churchland's eliminativism, the proposal to replace beliefs and desires by neurospeak, has taken a new turn in the guise of New Wave Reductionism and new-wave metascience (Bickle, 1998, 2003, 2006). John Bickle (2006) has claimed that we now have reductive explanations for such psychological phenomena as memory, attention, and consciousness in terms of cellular or even molecular neural processes (in some cases we even have the complete molecule-to-mind link, he asserts), so that we are now a step closer to a biophysics of the mind.

So there are roughly two approaches to naturalizing folk psychology: vindication and elimination. Either common-sense belief-desire psychology is salvaged by its incorporation in cognitive science, or we abandon appealing to beliefs and desires, and we forget about the mental realm as a really existing entity. In the debate on connectionism and folk psychology, eliminativists (Churchland, 1981, 1991) seem to assume that connectionism will replace folk psychology (see Ramsey, Stich, & Garon, 1991). To the extent that neural networks are a good simulation of our cognitive life, we do not have beliefs and desires, and explaining our fellow beings as caused by such inner springs cannot be literally correct.

However, it can be argued that these two approaches are not exhaustive, and that there is a way of keeping cognitive and neuroscience apart from common-sense psychology. In this view, neuroscience and folk psychology just have different job descriptions (Bechtel & Abrahamsen, 1993): folk psychology gives a coarse-grained account of global adaptive behavior, whereas neuroscience gives a fine-grained account of the underlying mechanisms. Folk psychology is about *personal*-level global behavior, scientific psychology is about *subpersonal*-level underlying mechanisms (Clark, 1996; Dennett, 1978). The latter roughly describes what rational purposeful people can and should do in a normal environment, not what the inner structure and causes of that behavior are. So, on this view, the big question, whether folk psychology will be replaced by cognitive and/or neuroscience, dissolves into a pluralism of explanations with different levels of description and prediction. It seems unlikely that the idealizing level of purposeful action can be completely captured in computational or neural terminology.

In stark contrast to the naturalistic framework sketched above, continental philosophy has emphasized the dichotomy between explanation in science and understanding (*Verstehen*) in the humanities. Since Brentano, having mental content is seen as the mark of the mental, rendering the mental distinct from the physical world (including the brain). Both Brentano (1874) and Dilthey (1894/1961) were much interested in keeping natural sciences away from mind, meaning, and culture. One approach to folk psychology, in the tradition of Dilthey and Brentano, is to explore the connection between empathy and the notion of understanding, as presumably distinct from scientific explanation (Heal, 2003). Intuitively, the hermeneutic endeavor of re-creating experience and unveiling hidden meanings of cultural artifacts must have something to do with mind reading, the capacity to understand others. Seen this way, the hermeneutic line of thought thus militates against naturalizing (and reducing) understanding. We will return to hermeneutics and its implications for mind reading below.

3. Psychological Theories of Mind Reading

3.1 Mind Reading: Theory or Simulation?

In developmental psychology, the empirical question how exactly the skills of empathy and prediction ("mind reading") work has attracted much attention.

Two classes of theories have been proposed for the psychological mechanism of mind reading: the first sees mind reading as theory-based, the second as a kind of simulation. On the first, received view (the theory theory), the beliefs, desires, and intentions of other people are inferred by applying an information-rich theoretical framework; on the second (the simulation theory), on the second one uses one's own decision mechanisms as a model of the other's mind (Nichols & Stich, 2003).

The theory theory holds that judging behavior in terms of beliefs and desires is applying theoretical concepts which refer to (presumed) laws; such laws explain acts as caused by mental contents. This is not to say that application of such generalizations occurs explicitly or that it must be conscious; on the contrary, we know these theoretical principles only tacitly and implicitly. Nor is it assumed that these laws are as interesting as new discoveries in physics – when put in words, they mostly sound like platitudes. As Nichols and Stich (2003) argue, it is very plausible that the theory of mind that underlies empathy is richer than the platitudes that we could consciously come up with.

It is a matter of ongoing dispute whether the theory of mind is acquired through learning during childhood or whether it is a modular phenomenon (which may or may not be given at birth). The former view holds that a body of theoretical explanations that appeal to internal beliefs and desires gradually develops through hypothesizing, testing and revising: the child is like a little scientist, developing a theoretical taxonomy of the mind. The latter view asserts that children have an innate, encapsulated, domain-specific, gradually maturing module (a "Theory of Mind Mechanism" (ToMM) – Leslie, 1994) that is specific for the domain of social cognition and is specialized in fast and dedicated computation that enables a person to infer beliefs and desires. Presumably, autists lack such a specialized neural mechanism, although they have roughly the same cognitive powers as others (Baron-Cohen, 1995).

An alternative view is that folk psychology might be a kind of simulation; we might just put ourselves in other people's shoes and imagine, "simulate," what we would do, think, or feel ourselves if we were in their situation. This is a kind of imaginative or dramatic skill, not the application of a *theory* of the laws of behavior. A useful analogy is the following: the simulation theory is like building a scale model of (for example) an airplane and see how it behaves in the air (Heal, 2003). The theory theory, in contrast, is like trying to predict its behavior according to the laws of gravity and aerodynamics. In the former case, the "model" is our own decision-making mechanisms and we run these to simulate the "target" (the other, whose mind we are trying to read). According to the simulation theory, "pretend beliefs" that the target is hypothesized to have are fed into these mechanisms, and the output of the belief generator is sent "off-line": the resulting belief is not acted upon but stored in a belief-predicting system. So, you use your own decision-making as a model to simulate the target, more or less like the model airplane is used to simulate the real one. The point is that you don't need theories or common-sense laws, as the theory theory would require, just your own cognitive mechanisms, fed with knowledge of the situation your target is in.

Both theory theory and simulation theory have spawned much empirical research and discussion; solving the debate is beyond the scope of this chapter, but plausibly, mind reading is a multifaceted skill, which requires both theory and simulation-like components (Nichols & Stich, 2003).

The standard test for mind reading is the so-called false-belief test, where the subject watches how the protagonist (a puppet) is led to believe that (for example) a box contains chocolate, which has however been removed while the protagonist is not looking (but the subject is). Children below four usually fail to recognize that other people may have beliefs that may be at odds with reality, and predict that the protagonist will not look in the box for the chocolate – note that the child in that case does not realize that the protagonist has a false belief. Depending on which of the theories one prefers, this can be explained either by a lack of theoretical tools, or by a lack of the dramatic skill to put oneself in the other's shoes (simulate), or by a yet immature module for mind reading.

The theory theory holds that a body of theoretical explanations that appeal to internal beliefs and desires gradually develops through hypothesizing, testing, and revising: the child is like a little scientist, developing a theoretical taxonomy of the mind. It is not very clear however, where the initial (unobservable) concepts for mental states come from: apparently, they are postulated somehow. In contrast, the simulation theory does have something to say about the origin of mental state concepts: these derive from one's own phenomenal experience. The child is initially directly aware of its own experience, since that is available when simulating, and therefore does not have to postulate theoretical constructs (Ohreen, 2004; Astington, 1996). (For a sophisticated account of the stage-wise development of understanding other minds from a basic imitative connection between the child's own acts and those of its caregivers, see Meltzoff's (2005) "like-me" theory.)

3.2 A Third Theory: The Sociolinguistic View

The traditional theories of folk psychological competence all assume that the belief-desire explanations are universal throughout different cultures. The finding that belief-desire psychology is not universal (Lillard, 1998) is therefore a strong empirical reason to doubt these theories. Ethnopsychology shows that there are many alternative ways of explaining behavior that do not appeal to mental states as its causes. For example, animism attributes emotions to sources outside the person; Homer's heroes experienced their actions as triggered by the gods; and in less individualist societies than ours conformity to group norms rather than individual mental states explains behavior. The Western concept of mind as a container of causally effective beliefs and desires and emotions seems absent or less prominent in other cultures. As not all cultures have the same theory of mind, cultural influence should be taken into account in explaining the ToM, and mind reading is not the operation of an innate universal module. Cultural variation is not easily explained by the simulation hypothesis either.

From this sociolinguistic perspective, Ohreen (2004) argues for the radical antirealist conclusion that beliefs and desires as characterized by Fodor do not really exist. Rather they are conceptual tools in a Wittgensteinian language game, and although they may be useful in one cultural form of life, other communities may use different linguistic tools.

Furthermore, he argues, understanding others as thinking and feeling beings is not necessarily the same as attributing beliefs and desires to them. Presumably, we have also a more primitive and biological level of understanding others, which does not rely on fully-fledged, explicit concepts of beliefs and desires (nor on simulation). The child starts from such a primitive understanding, which allows it to directly experience the other's emotional states. It could perhaps be compared to a kind of mental contagion, for example when one feels sad when seeing others cry. Following Tomasello (see below), Ohreen assumes a basic biological understanding of others, on which linguistic and cultural learning is built. Culture transforms the basic biological processes into new and more sophisticated ways of thinking.

Therefore, the false-belief task is too limited a test of social cognition. Young children do have some basic understanding of other minds, even when failing in the attribution of discrete beliefs and desires.

Recognizing others as thinking and feeling does not necessarily require attributing the explicit, symbolic mental states that the computational theory assumes. Empathy may consist in (causally effective) mental states that resonate somehow to other people's feelings, but not necessarily involve recognizing distinct beliefs, nor correct answers in the false-belief task.

In the sociolinguistic view, folk psychology is a matter of concept acquisition, which makes it a cultural rather than individual mental phenomenon. It is a matter of learning to use a certain discourse in a social and cultural context. Mental concepts as inner causes of behavior is a Western invention, other cultures may have different conceptual tools. Folk psychology is created by culture, as such it is an artifact like houses and swimming pools (Ohreen, 2004, p. 72). Astington's (1996) and Ohreen's (2004) sociocultural theory is explicitly Vygotskyan: mind is internalized from the outside, it develops by taking part in the cultural way of talking about other peoples' behavior, rather than as an innate module, or a proto-scientific set of abstract generalizations. A first, preliminary conclusion then is that explaining how we understand other agents seems to require more than just internal mechanisms.

4. An Antireductionist Interpretation of Simulation: *Verstehen*

In nineteenth-century continental philosophy meaning, consciousness, feeling, experience, and rationality were considered inaccessible to hard science. The domain of hermeneutics that was concerned with understanding those higher properties of humans (*Verstehen*) was set sharply apart from explanation in the sciences (*Erklären*). The Cognitive Revolution of the 1960s can be seen as an attempt to make the elusive mental properties amenable to objective scientific explanation (Gardner, 1985) by construing mental states as computational states of an (abstract) machine. Both classical symbolic models and connectionism are projects to naturalize the mind, showing how rationality is mechanically possible (Fodor, 1987, p. 20) and how folk psychology could be legitimized and explained within a computational model of mental processes. Mind reading or empathy is of course closely related to hermeneutic understanding, and if cognitive psychology can provide a mechanistic explanation in terms of information processing mechanisms, the naturalization of such higher mental processes as mind reading lies around the corner.

The simulation theory emphasizes the imaginative or dramatic skill to impersonate the mental life of the target as the basis of mind reading. Interestingly, such a capacity to put oneself in the other's shoes looks much like the subjective prejudice that the hermeneuticist brings to understanding (Gadamer, 1975). Thus, traditional hermeneutics suggests itself as an alternative to the naturalistic cognitive approach to simulation. Cognitivists (Gallese & Goldman, 1998; Nichols & Stich, 2003) reconstruct simulation in terms of decision mechanisms and "off-line beliefs", within the framework of mainstream cognitive science – an engineering approach, using a "boxology" of flow diagrams and information processing subroutines (see Nichols & Stich, 2003).

The alternative hermeneutic approach objects that the naturalist mainstream mistakenly tries to assimilate psychology, in particular beliefs, desires, and empathy, to the model of the hard sciences. According to hermeneuticists, the faculty of imagination providing access to other people's minds, making judgments about others' thoughts, and judging their rationality, does not fit in the natural science model. Jane Heal (2003) proposes a view of simulation (which she also calls reproduction or co-cognition) that is congenial to the *Verstehen* tradition. Treating the other as a rational and responsible person is different from the objective scientific method that focuses upon the causal role of mental states. Heal's alternative, like *Verstehen*, is that we recreate from our own subjectivity the state of mind of the other. It involves thinking about the same reality, the same subject matter, that the target thinks about, with (as far as possible) the same reasoning abilities and background beliefs. The term "co-cognition" is coined for this variety of mind reading.

So, the "cognitive" approach to simulation engages in empirical investigations trying to specify the cognitive machinery that produces psychological judgments (e.g., Gallese & Goldman, 1998). Heal's "hermeneutic"[1] variety does not consider one's fellow beings as machines whose internal mechanisms one must know and manipulate to predict and control. Rather, according to this approach one tries to understand from the inside, from our own person, recreating the way others construct their view of the world. Thus, the hermeneutic version of mind reading (co-cognition) is a personal-level view, whereas the cognitivist version (the boxology of off-line simulation etc.) can be seen as a sub-personal claim about the way co-cognition is realized. This intersubjective principle is missing in the subpersonal cognitivist view. We can debate on the correctness of our reasoning against a background practice that we can and should give reasons and ask the other for reasons, in a way that formal (computational) systems cannot, Heal argues.

If we accept that argument, mind reading is more than cognitive (let alone neural) machinery. As we have seen, the sociolinguistic criticism of the naturalistic cognitive theories of folk psychology implies that the explanation of mind reading lies beyond the individual mind/brain.

So, the consequence of the sociolinguistic approach to folk psychology and Heal's hermeneutic construal of simulation seems to be that the naturalization of mind reading in its cultural manifestations is a lost cause. The hermeneutic approach implies a chasm between the personal and the subpersonal, between rational understanding and cognitive mechanisms, between culture and biology. Hermeneutics deliberately creates a dichotomy between the subpersonal mechanisms that are the subject of a naturalistic, reductionist approach (*Erklären*), and the personal-level recreation of subjective meaning (*Verstehen*). The former has its domain on the natural sciences (including biology) (*Naturwissenschaften*), the latter in the humanities and cultural studies (*Geisteswissenschaften*) – and never the twain shall meet.

5. Biology and Culture in Mind Reading

The hermeneutic approach thus seems to cut off higher mental capacities from the cognitive and neurosciences as it draws a Dilthey-style dichotomy between the personal-level capacity of sharing intentions and experiences on the one hand, and the subpersonal level of cognitive and biological mechanisms on the other hand. This is clearly an unsatisfactory account of explanatory practices in cognitive science: there is interesting empirical work on the underlying mechanism of shared cognition and goals, suggesting how to connect biological and cultural influences on development of mind. The dichotomy that seemed to follow from our discussion above is surely simplistic.

Tomasello (1999) presented a view of human cognition that spans the biological and the cultural, and answers some questions on the origin of mental concepts; importantly, in his view, both biological and cultural inheritance is required in the development of empathy. The basic idea is that humans have a unique biological adaptation for social cognition. Understanding others as intentional agents enables both sociogenesis, creating culturally based tools and collective cognitive skills, and internalization of this collective tool kit into individual capacities.

The capacity for understanding the mental states underlying behavior is probably not entirely unique to humans: initially Tomasello (1999, pp. 15–19) assumed that other primates do understand something about manipulating external relations, but not about internal causes that are not directly observable. This was thought to apply to causality as well as to mentality: primates do not understand intentionality, the beliefs and desires behind behavior, and they do not understand hidden causes. They do understand some form of manipulation in the physical realm (shaking trees to get fruit), but seem to lack the idea of an invisible physical cause, and likewise the idea of an intention. The "why" behind the effects, the mediating processes underlying the observed external relations escape nonhuman primates. Tomasello's hypothesis is that understanding other agents' behavior in terms of intentions was phylogenetically first, and was then carried over to inanimate objects in the guise of underlying causes. So, intentional and causal thought were supposedly uniquely human capacities. Recent evidence however indicates that great apes and children with autism seem to have some understanding of intentionality: they recognize goals in others and understand that others perceive things (they follow gaze) (Tomasello et al., 2005), but they do not seem to want to share these insights with their conspecifics. Apes show some coordinated behavior, but do not really collaborate towards shared goals nor attempt to share interests and experiences. Showing others the use and meaning of things is not in their repertoire, apparently. The same seems true of children with autism.

The ability to understand others as intentional agents, and detecting their communicative intentions is the basis for understanding gestures and signs, and ultimately for learning language. Internalizing language and other conventional symbols requires understanding communicative intentions, in what Tomasello calls a joint attention situation. Language acquisition (learning the conventional use of symbols) requires such a context, where the child (18–24 months) recognizes the other as an intentional agent and determines his communicative intentions. Tomasello emphasizes that cognitive development is interpersonal and perspectival: the culturally produced ways of seeing the world through concepts are internalized by the individual in a joint attention setting: "Understanding other persons as intentional agents makes possible both (a) processes of sociogenesis by which multiple individuals collaboratively create cultural artifacts with accumulative histories, and (b) processes of cultural learning and internalization by means of which developing individuals learn to use and then internalize aspects of the collaborative products created by conspecifics" (Tomasello, 1999, p. 15).

Explicit belief/desire attributions, that traditionally are taken as basic in the folk psychology debate, are in Tomasello's view derived from shared intentionality. This seems in accordance with Ohreen's thesis that the false-belief test is not necessarily an adequate measure of social cognition: joint attention, detecting communicative intentions may be possible without protoscientific generalizations based on attributing discrete beliefs and desires.

Shared intentionality thus is the basis of culture: joint attention explains typical human achievements as language and mind reading. As in Vygotsky, internalizing is an important part of Tomasello's theory: joint attention is internalized into symbolic representations (1999, p. 131). The symbol allows new perspectives on the world, and enables new courses of action – cultural beings, unlike primates, have a flexibility that the fixed sensory-motor representations in primates lack.

Obviously, the manifold and complicated issues of interaction between biological adaptation and culture are beyond the scope of this chapter (see e.g., Laland & Brown, 2002, Sperber & Hirschfeld, 2004). Important issues like the role of imitation in mind reading, the nature of the first-person to third-person analogy (the "like me") (see Goldman, 2005; Gordon, 2005; Hurley & Chater, 2005), and so on, are not yet resolved.

The point we want to emphasize is that in understanding mind reading, a multilevel view is indispensable. Tomasello (1999, p. 15) writes: "most, if not all, of the species-unique skills of human beings are not due to a unique biological inheritance directly, but rather result from a variety of historical and ontogenetic processes that are set into motion by the one uniquely human, biologically inherited, cognitive capacity."

What this very brief discussion of the cultural origins of cognition shows is that mind reading is part of a biological-cultural-developmental network. Distinctive human capacities like language, culture, symbolic perspectives, imitation, communication and mind reading are ontogenetically and phylogenetically interwoven. On the one hand, the story of social cognition cannot be told from a culture-free perspective; on the other hand it cannot be told ignoring developmental psychology, evolutionary biology and ethology. As we saw before in the socio-cultural or sociolinguistic theory of folk psychology, the individualist tendency of the simulation and theory theories needs correction. The data suggest that belief-desire psychology cannot be considered in isolation, and no equipotent image, no complete explanation of empathy can be built bottom-up from subpersonal mechanisms.

However, recent developments in neuroscience suggest that a direct link between neural processes and empathy can be established; if true, such a link would support the eliminativist project, explaining mind reading as neural processes.

6. Understanding Neurons?

6.1 Mirror Neurons

Recently, claims have been put forward that so-called mirror neurons are the basis of understanding others' intentions and experiences (Rizzolatti & Craighero, 2004). Abundant evidence now strongly suggests that certain visuomotor neurons in the monkey ventral premotor cortex area F5 (and a few other brain centers, such as STS (Superior Temporal Sulcus) and PF in the inferior parietal lobe) not only discharge when the monkey performs a goal-directed action like holding, manipulating or grasping, but also when the monkey observes the same or similar object-directed actions performed by another individual, either a conspecific or a human experimenter. For instance, grasping a banana and seeing an experimenter grasp it evoke similar activation patterns. Mirror neurons respond to an action independently of value or reward, and the response generalizes over a widely diverse range of visual stimuli, as long as the same action is involved. They do not respond when the monkey observes a hand miming an action or when confronted with the sight of an isolated object (Umiltà et al., 2001). Neither do the activation patterns differentiate between actions that take place near or far from the observing monkey. Nor is it necessary that the monkey actually sees the action: understanding an action on the basis of sound (like breaking peanuts) is effective as well. So the evidence indicates that it is the meaning, not the visual features of the action that are represented in mirror neurons (Rizzolatti & Craighero, 2004). Evidence for a similar system in humans, coding for both observed and executed actions, comes from EEG, brain imaging studies (see Gallagher & Frith, 2003) and TMS (Transcranial Magnetic Stimulation) (single neuron recordings in humans are rarely possible of course). Most researchers seem to consider the neural activation measured by brain imaging techniques in the inferior frontal cortex in humans as driven by mirror neurons. The human motor-neuron system seems to resonate to a wider range of properties than in monkeys: the human mirror system, in contrast to the monkey mirror system, for instance responds to mimed action (Gallese, Keyzers, & Rizzolatti, 2004).

According to the discoverer of the motor neuron system, Giacomo Rizzolatti, observing an action activates the same neural network that the observer uses in executing that action. This activation amounts to understanding that action – he understands the action because he knows the outcome when he does it (Gallese et al., 2004, p. 396); observing an action triggers the first-person motor knowledge of the observer. So, Gallese et al. claim that we have a mechanism for direct experiential insight in other minds, including the understanding of emotion (Gallese et al., 2004, p. 402). (The latter is presumably located in the insula.) Brain-imaging studies suggest that in imitation learning in humans the same mirror neuron areas are used as in action understanding. So, the interesting suggestion is that activation of motor representation amounts to understanding an action because it triggers a kind of "as if" execution of the same action. Mirror neurons give a kind of direct communicative resonance between actors, bridging first person and third person knowledge (cf. Decety and Chaminade, 2003). This as-if experience seems an obvious basis for empathy, understanding and *Verstehen*; it is apparently a mechanism that allows us a direct experiential grasp of the mind of others (Gallese et al., 2004).

6.2 A Unifying Neurophysiological Theory of Understanding?

According to Gallese and Goldman (1998), the simulation theory of folk psychology seems supported by the mirror neuron findings. They describe the function of mirror neurons as a system that matches observation and execution of motor actions, and suggest that mirror neurons are the basis for an evolutionary precursor of "mind reading": they may serve to reason backwards from movement to (simulated) intention. Mirror neurons not only encode one's own internally generated action plan, they also reflect an (externally generated) recreation of the intention of the other; i.e. the system creates a state in the observer that matches the motor plan of the target. According to Gallese and Goldman, this looks like a primitive form of mental simulation: mirror neurons, they say, appear to be "nature's way of getting the observer into the same 'mental shoes' as the target – exactly what the conjectured simulation heuristic aims to do" (pp. 497–498). Perhaps we have here a brain mechanism for understanding action-goals that in the course of evolution developed into empathy and mind reading abilities. Of course, more would be needed to refute the hermeneuticist's co-cognition version of the simulation theory (Heal, 2003) and the criticism that subpersonal mechanisms cannot capture the full personal-level intersubjective process of co-cognition. It does call attention, however, to the question how co-cognition fits in with evolutionary lower mechanisms of mind reading. As we saw above, the dichotomy of explaining and understanding, and the isolation of personal from subpersonal level, separating the subjective and cultural from biological processes, that some antireductionists build their case on, is not an attractive option.

This is, then, arguably a test case for reduction: understanding the meaning of action (as distinct from mere external behavior), empathy, and shared experience might have a distinct neural locus and mechanism. A neuroscience of mind reading seems around the corner. This is apparently what Gallese et al. (2004) have in mind, when they claim that the mirror neuron system provides a unifying perspective of the neural basis of social cognition (p. 401); that this neural mechanism allows us to directly understand the meaning of the actions of others by internally replicating (simulating) them (p. 396); and that we now have a neurophysiological account of the experiential dimension of action and emotion (p. 396).

6.3 Mirror Neurons Code for Intentions – As Seen from the Context

In our opinion, Gallese et al.'s (2004) claim is not warranted. Jacob and Jeannerod (2005) point out that the mirror neuron system is limited to (object-directed) motor actions. Communicative intentions do not necessarily map onto movements: the same movement can be associated with different social intentions. For instance, the same movement of pointing at one's watch may communicate that the party is over, that the watch is broken, that one is wearing a birthday present, or whatever. Simulating and understanding a movement thus is not sufficient for recognizing, for instance, communicative intentions. Moreover, Jacob and Jeannerod (2005) point out that the STS, the amygdala, the orbitofrontal cortex and the inferior parietal lobule together constitute a social perception system, but that these have no motor resonance. So, at the very least not all simulation is mirroring. (That does not necessarily affect the status of simulation theory – its conceptual and empirical foundation is not necessarily tied to a motor hypothesis).[2] Therefore, not all social perception can be explained in terms of motor mirroring, Jacob and Jeannerod argue.

Recently however, a strong case for localizing understanding in mirror neurons has been made by Iacoboni et al. (2005).[3] They investigated the claim whether the mirror neuron mechanism is primarily for action recognition or for action understanding (or intention understanding), that is, do mirror neurons code only for *what* is being done (and how), or also for *why* it is being done? A condition in which only a scene was shown (tea cups ready either for tea drinking or for cleaning) was compared with a condition in which only the act of grasping the cup was shown, and with a third condition in which the cup-grasping was shown in the context of either drinking or cleaning. Watching cup-grasping gives the expected activation in the inferior frontal area of the cortex where the hand motor action is represented (measured by fMRI in humans.); this was absent when watching the scene without action. Presenting the action in context (drinking or cleaning) evokes increased activation, as compared to watching the action in isolation. So, it makes a difference at the mirror neuron level whether just the act of grasping a cup is seen, or whether it seen as done with the intention to drink from it or to wash it.

Iacoboni et al. conclude that the mirror neuron system not only reflects action recognition, but is also involved in understanding the intention of others – not only what the other does, but also what he intends to do. This is an important finding, since it connects the mirror neuron system to higher level intentions. Furthermore, they contend that inferring a goal is an automatic operation, because no differences were found in the right inferior frontal area between conditions in which subjects were attending and when they were not.

Also, there is a difference in activation between the drinking and the cleaning scenes in the intention condition (but not in the other two conditions), suggesting that the difference in the respective goals of the observed action is reflected in differential neural activation. Iacoboni et al. hypothesize that a set of mirror neurons other than the classical mirror neurons, or a subset of them, computes a possible subsequent action on the basis of the context; drinking is a more common and basic activity and therefore recruits more neurons than cleaning. Perhaps, they suggest, detecting an intention *just is* detecting a new goal.

Is this corroboration for the reductionist view, directly locating intention in neurons? Not necessarily, we think. As Iacoboni et al. note, the intention-detecting neurons are presumably different from, or a subset of, the classical mirror neurons. That suggests that the neural basis of social cognition is an even more differentiated system than Jacob and Jeannerod suggested. Also, as Iacoboni et al. hypothesize, the neurons may not code for the underlying intention proper, but for an extrapolation of the subsequent motor action.

Most importantly however, taking account of the context in which mirror neuron system works, i.e. its connections with the rest of the nervous system, the task the organism is executing, the environment, the organism's learning history, etcetera, is still needed to make sense of a neural process. It is obvious from Iacoboni et al.'s paper that their interpretation of fMRI activation patterns involves interpreting the processing demands imposed by the experimental conditions, and that learning history (presumably explaining why drinking action is neurally different from cleaning action) is part and parcel of the explanation of the results.

As the New Mechanicists (cf. Craver, 2001; Bechtel & Abrahamsen, 2005; Bechtel, Chapter 8 in this volume) have emphasized, "looking up for context" is an indispensable part of the explanation of a mechanism. In our case, systems neuroscience, environment, leaning history and task demands enter into the interpretation and explanation of mirror neuron activation. That implies that the straight arrow from below, explaining psychological processes directly from neural processes is not going to work. The bottom-up story ignoring context cannot be complete. Although it is too early for a firm conclusion, it seems fair to say on the basis of current evidence that he mirror neuron system provides no unifying neural explanation, no reductive equipotent image of mind reading.

6.4 Development of Mind Reading, Gesturing, and Language: Multiple Connections

As we have seen, understanding the neural basis of a psychological process like empathy is more complicated than a direct identification of neural activation with understanding; the mirror neurons are part of a larger system with complex interconnections. Looking up for context (Craver, 2001), at the way the mirror neuron system is embedded in the organism and the function it serves, at the way the organism is embedded in the environment and the task it is accomplishing is an indispensable part of explaining neural states and processes.

However, in the case of mind reading the picture seems even more complicated than the mechanicists envisage. As we saw above, the mirror neuron system may be related to sharing intentions in a joint attention situation. That requires not only looking up for context, but also looking back in phylogenetical and ontogenetical history. A further interesting connection is how the mirror neuron system might be related to language. Shared attention is crucial for cooperation, for solving coordination problems by directing the other's attention; language seems a natural extension of this system. It seems possible that language developed from the basic mirror neuron system for grasping, through a system for gesturing. The picture is very roughly that a more or less straight evolutionary line runs from the mechanisms of mirror neurons underlying mind reading, to proto-signs or gestures, to proto-speech sounds, and finally to fully fledged human speech and language (Arbib, 2005; Corballis, 2003). The structure of complex social acts may be a step on the evolutionary ladder towards linguistic competence. Interestingly, the homologue in primates of the Broca area in humans is the ventral premotor area (F5) – which happens to be the location of the primate mirror neuron system. In monkeys this area has nothing to do with vocal control. This supports the hypothesis that vocalization has developed from gesturing ("from hand to mouth," as Corballis, 2003 put it).[4]

So, whatever the final contours of a theory of the development of mind reading, communication and language will be, the brief and speculative remarks above suggest that language, gesturing, shared intentionality and empathy are intertwined. To explain language and communication we need a patchwork of hypotheses weaving together culture, the brain, cooperation, empathy, tool use, imitation, brain mechanisms of motor recognition, brain mechanisms of shared intention. This patchwork will span several levels of organization (including neural and mental mechanisms, evolution, society, and culture). Exploring these interconnections requires looking upward, to contexts at higher levels of organization, sideward to other domains of psychology (language, development, cooperation), and backward in evolution and ontogeny. The result is not only a multilevel, but also a thoroughly pluralist picture of social cognition and communicative intent.

7. Conclusion: Multiple Perspectives on Mind Reading

The deeper question behind the case of mirror neurons was whether psychology should take its basic concepts from natural sciences, and aim for a reductive account of rationality, understanding and mind reading. One line of thought is that a science in terms of mind, mental content, rationality, and intentionality is an obsolete and confused way of talking, a philosophers' fine conceptual mess that is best eliminated and forgotten (Bickle, 1998; Churchland, 2002). Some neurophilosophers would opt for a straight arrow of reduction from cognition to molecules (Bickle, 2003, 2006), others are more circumspect about the idea of "explanation in a single bound" (Churchland, 2002, p. 29).

The other extreme is that empathy is beyond the reach of empirical science and lies at the unique irreducible personal level (Heal, 2003): understanding other agents' intentions requires an irreducibly holist, normative perspective that is incompatible with a reductive approach. As mentioned above, Heal (2003) sheds some doubts on the mechanistic subpersonal concept of off-line processing as an account of empathy. Her arguments touch upon some deep issues in hermeneutics:

naturalism and the mechanisms of cognition and nervous system cannot do justice to co-cognition or shared rationality.

In our opinion the most plausible view moves beyond these two extremes and recognizes that there is a multiplicity of ways of understanding human behavior, and that these serve different explanatory purposes. Personal/intentional and subpersonal/mechanical explanations can not only coexist alongside each other, but also mutually correct and enrich each other (cf. McCauley, 1996; Bechtel & Abrahamsen, 1993). We encounter here again the pluralism of explanations that (dis)solved the folk psychology debate (see Section 2). A personal-level approach like hermeneutics is in this view just a perspective among others, which has its distinct explanatory uses; however, it should not be taken as absolute and used to discard subpersonal explanations. So we should reject both the exaggerated interpretation of mirror neurons as a complete explanation of social cognition, and the dichotomy created by philosophical hermeneutics between understanding and explanation, and between a personal and subpersonal level. Rather, we believe that explanations at the levels of neuroscience, evolutionary biology, personal-level constructs, evolutionary psychology, and cognitive psychology can fruitfully interact.

In the research mentioned above we have seen that mind reading cannot be isolated from other capacities, such as language on the one hand, and primitive mechanisms for detecting intentions on the other hand. Some evidence suggests that language is intertwined with gesturing and with detecting communicative intent. All these are potentially fruitful inter theoretical and interlevel connections that would be lost both in eliminativist and in dualist approaches. Against the eliminativist, it should be noted that the higher level is not just heuristics, not just setting op the explanandum for neuroscience (cf. Bickle, 1998); folk psychology has its explanatory uses in some contexts. Likewise, theorizing on the functions and evolutionary origins of imitation, gesturing, language is a *bona fide* constraint on interpretations of the role of neural mechanisms (Schouten & Looren de Jong, 1999; Looren de Jong, 2002). Mirror neurons are embedded in a biological-cultural context and can only be understood within that context.

To sum up then, there is no simple reduction of mind reading to, or elimination by, mirror neurons. We see a network of explanations crossing levels, not a straight arrow of reduction down to molecular neuroscience, as the eliminativist would have it. No equipotent image of the explanations and constraints from common sense, evolutionary theory, ethology, neuronatomy, evolutionary homology, linguistics, and so on, is available as a bottom-up construction from basic neuroscience. The situation in social neuroscience can probably be best characterized as coevolution of theories at different levels (Looren de Jong, 2002; McCauley, 1996), mutual adjustment of concepts and theories (Schouten & Looren de Jong, 1998; van Eck, Looren de Jong & Schouten, 2006), and explanatory pluralism.

Acknowledgments

We wish to thank Bill Bechtel and John Bickle for their criticisms and suggestions on an earlier draft of this chapter.

Notes

1 Continental hermeneutics is somewhat deeper than the account given here. In particular, in the later Dilthey, and in Gadamer, *Verstehen* is about interpreting the meaning of objectifications of the objective spirit, rather than on recreating subjective feeling. Let us consider Heal's variety of hermeneutics as a kind of "hermeneutics lite."

2 Wicker et al. (2003) is a good example of an automatic view of "mental contagion": they found the same site in the anterior insula is activated when subjects smell a disgusting odor, as when they see another person look disgusted. The hypothesis is that the insula is a common substrate for both feeling and perceiving emotion in others.

3 Thanks to Helen De Cruz for pointing this out.

4 Floel et al. (2003) found that language perception (bilaterally) activates the hand motor system (Transcranial Magnetic Stimulation elicited larger electromyographic responses in speech production and speech perception tasks than in nonverbal tasks). They speculate that communicative gestures are coded in an action-perception network; therefore, language perception can be expected to activate hand motor system (but not other systems, and nonlinguistic tasks will not activate the motor system). This activation is bi-lateral, whereas speech areas in humans are usually localized in the left hemisphere. This suggests that language evolved from manual gesturing, and that specialized (lateralized) language understanding builds upon a phylogenetically older (bilateral) sign system (Floel et al., 2003).

References

Arbib, M. A. (2005). From monkey-like action recognition to human language: An evolutionary framework for neurolinguistics. *Behavioral and Brain Sciences*, *28*, 105–167.

Astington, J. (1996). What is theoretical about the child's theory of mind? In P. Carruthers & P. K. Smith (Eds.), *Theories of Theories of Mind* (pp. 185–199). Cambridge: Cambridge University Press.

Baron-Cohen, S. (1995). *Mind-Blindness*. Cambridge, MA: MIT Press.

Bechtel, W., & Abrahamsen, A. A. (1993). Connectionism and the future of folk psychology. In S. M. Christensen & D. R. Turner (Eds.), *Folk Psychology and the Philosophy of Mind* (pp. 340–367). Hillsdale, NJ: Lawrence Erlbaum, and in R. Burton (Ed.), *Minds: Natural and Artificial* (pp. 69–100). Albany, NY: SUNY Press.

Bechtel, W., & Abrahamsen, A. (2005). Explanation: A mechanistic alternative. *Studies in History and Philosophy of the Biological and Biomedical Sciences*, *36*, 421–441.

Bickle, J. (1998). *Psychoneural Reduction: The New Wave*. Cambridge, MA: MIT Press.

Bickle, J. (2003). *Philosophy and Neuroscience: A Ruthlessly Reductive Account*. Dordrecht: Kluwer Academic Publishers.

Bickle, J. (2006). Reducing mind to molecular pathways: Explicating the reductionism implicit in "molecular and cellular cognition." *Synthese*, forthcoming.

Brentano, F. (1874). *Psychologie vom empirischen Standpunkt* (2nd, enl. ed. by Oskar Kraus, 1924–1925). Hamburg: Meiner Verlag.

Churchland, P. M. (1981). Eliminative materialism and the propositional attitudes. *Journal of Philosophy*, *78*, 67–90.

Churchland, P. M. (1989). *A Neurocomputational Perspective: The Nature of Mind and the Structure of Science*. Cambridge, MA: MIT Press.

Churchland, P. M. (1991). Folk psychology and the explanation of human behavior. In R. J. Bogdan (Ed.), *Mind and Common Sense* (pp. 37–52). Cambridge: Cambridge University Press.

Churchland, P. S. (2002). *Brain-Wise: Studies in Neurophilosophy*. Cambridge, MA: MIT Press.

Clark, A. (1996). Dealing in futures. In R. N. McCauley (Ed.), *The Churchlands and Their Critics* (pp. 86–103). Oxford: Blackwell.

Corballis, M. C. (2003). From mouth to hand: Gesture, speech, and the evolution of right-handedness. *Behavioral and Brain Sciences*, *26*, 199–260.

Craver, C. F. (2001). Role functions, mechanisms and hierarchy. *Philosophy of Science*, *68*, 53–74.

Decety, J., & Chaminade, T. (2003). When the self represents the other: A new cognitive neuroscience view on psychological identification. *Consciousness and Cognition*, *12*, 577–596.

Dennett, D. C. (1978). Intentional systems. In D. C. Dennett, *Brainstorms* (pp. 3–22). Hassocks: Harvester.

Dilthey, W. (1894/1961). Ideen ueber eine beschreibende und zergliedernde Psychologie. In O. F. Bollnow (Ed.), *W. Dilthey: Die Philosophie des Lebens* (pp. 131–229). Stuttgart: Teubner.

Floel, A., Ellger, T., Breitenstein, C., & Knecht, S. (2003). Language perception activates the hand motor cortex: Implications for motor theories of speech perception. *European Journal of Neuroscience*, *18*, 704–708.

Fodor, J. A. (1981). *Representations: Philosophical Essays on the Foundations of Cognitive Science.* Hassocks: Harvester.

Fodor, J. A. (1987). *Psychosemantics: The Problem of Meaning in the Philosophy of Mind.* Cambridge, MA: MIT Press.

Fodor, J. A. (1990). Making mind matter more. In J. A. Fodor, *A Theory of Content and Other Essays* (pp. 137–159). Cambridge, MA: MIT Press.

Fodor, J. A. (1995). A theory of the child's theory of mind. In M. Davies & T. Stone (Eds.), *Mental Simulation* (pp. 109–22). Oxford: Blackwell.

Gadamer, H. G. (1975). *Wahrheit und Methode* (4. Auflage). Tübingen: Mohr.

Gallagher, H. L., & Frith, C. D. (2003). Functional imaging of "theory of mind." *Trends in Cognitive Sciences*, *7*, 77–83.

Gallese, V., & Goldman, A. (1998). Mirror neurons and the simulation theory of mind reading. *Trends in Cognitive Sciences*, *2*, 493–501.

Gallese, V., Keyzers, C., & Rizzolatti, G. (2004). A unifying view of the basis of human cognition. *Trends in Cognitive Sciences*, *8*, 397–403.

Gardner, H. (1985). *The Mind's New Science.* New York: Basic Books.

Goldman, A. I. (2005). Imitation, mind reading, and simulation. In S. Hurley & N. Chater (Eds.), *Perspectives on Imitation: From Neuroscience to Social Science. Vol 2: Imitation, Human Development, and Culture* (pp. 79–93). Cambridge, MA: MIT Press.

Gordon, R. I. (2005). Intentional agents like myself. In S. Hurley & N. Chater (Eds.), *Perspectives on Imitation: From Neuroscience to Social Science. Vol 2: Imitation, Human Development, and Culture* (pp. 95–106). Cambridge, MA: MIT Press.

Heal, J. (2003). *Mind, Reason and Imagination.* Cambridge: Cambridge University Press.

Hurley, S., & Chater, N. (2005). Introduction: The importance of imitation. In S. Hurley & N. Chater (Eds.), *Perspectives on Imitation: From Neuroscience to Social Science. Vol 2: Imitation, Human Development, and Culture* (pp. 1–52). Cambridge, MA: MIT Press.

Iacoboni, M., Molnar-Szakacs, I., Gallese, V., Buccino, G., & Mazziotta, J. C. (2005). Grasping the intentions of others with one's own mirror neuron system. *PLoS Biol*, *3(3)*, e79.

Jacob, P., & Jeannerod, M. (2005). The motor theory of social cognition: A critique. *Trends in Cognitive Sciences*, *9*, 21–25.

Laland, K. N., & Brown, G. R. (2002). *Sense and Nonsense: Evolutionary Perspectives on Human Behavior*. Oxford: Oxford University Press.

Leslie, A. (1994). ToMM, ToBY, and agency: Core architecture and domain specificity. In L. A. Hirschfeld & S. A. Gelman (Eds.), *Mapping the Mind: Domain Specificity in Cognition and Culture* (pp. 119–148). Cambridge: Cambridge University Press.

Lillard, A. (1998). Ethnopsychologies: Cultural variations in theories of mind. *Psychological Bulletin*, *123*, 3–32.

Looren de Jong, H. (2002). Levels of explanation in biological psychology. *Philosophical Psychology*, *15*, 441–462.

McCauley, R. N. (1996). Explanatory pluralism and the coevolution of theories in science. In R. N. McCauley (Ed.), *The Churchlands and Their Critics* (pp. 17–47). Oxford: Blackwell.

Meltzoff, A. N. (2005). Imitation and other minds. In S. Hurley & N. Chater (Eds.), *Perspectives on Imitation: From Neuroscience to Social Science. Vol 2: Imitation, Human Development, and Culture* (pp. 55–77). Cambridge, MA: MIT Press.

Nichols, S., & Stich, S. P. (2003). *Mindreading*. Oxford: Clarendon.

Ramsey, W., Stich, S., & Garon, J. (1990). Connectionism, eliminativism and the future of folk psychology. In J. E. Tomberlin (Ed.), *Philosophical Perspectives 4: Action Theory and Philosophy of Mind* (pp. 499–533). Atascadero, CA: Ridgeview.

Ohreen, D. E. (2004) *The Scope and Limits of Folk Psychology: A Sociolinguistic Approach*. Berne: Peter Lang.

Rizzolatti, G., & Craighero, L. (2004). The mirror-neuron system. *Annual Review of Neuroscience*, *27*, 169–192.

Schouten, M. K. D., & Looren de Jong, H. (1998). Defusing eliminative materialism: Reference and revision. *Philosophical Psychology*, *11*, 489–509.

Schouten, M. K. D., & Looren de Jong, H. (1999). Reduction, elimination, and levels: The case of the LTP-learning link. *Philosophical Psychology*, *12*, 239–264.

Sperber, D., & Hirschfeld, L. A. (2004). The cognitive foundations of cultural stability and diversity. *Trends in Cognitive Sciences*, *8*, 40–46.

Tomasello, M. (1999). *The Cultural Origins of Human Cognition*. Cambridge: Harvard University Press.

Tomasello, M., Carpenter, M., Call, J., Behne, T., & Moll, H. (2005). Understanding and sharing intentions: The origins of cultural cognition. *Behavioral and Brain Sciences*, *28*, 675–691.

Umiltà, M. A., Kohler, E., Gallese, V., Fogassi, L., Fadiga, L., Keysers, C., & Rizzolatti, G. (2001). I know what you are doing: A neurophysiological study. *Neuron*, *32*, 91–101.

van Eck, D., Looren de Jong, H., & Schouten, M. K. D. (2006). Reduction, elimination, and levels: The case of the LTP-learning link. *British Journal for the Philosophy of Science*, *57*, 167–196.

Wicker, B., Keysers, C., Plailly, J., Royet, J., Gallese, V., & Rizzolatti, G. (2003). Both of us disgusted in my insula: The common neural basis of seeing and feeling disgust. *Neuron*, *40*, 655–664.

NAME INDEX

SUBJECT INDEX